TROPICAL MOIST FORESTS

The Resource
The People
The Threat

Tropical moist forest (TMF) — "jungle", rainforest, cloudforest and swampforest - covers an area the size of the United States and is home to almost half the world's wild species.

How fast is TMF disappearing?

* At 5.6 million hectares a year, according to the most optimistic study;

* At 20 million hectares per year, according to the US National Academy of Sciences.

The UN, in an authoritative 1981 FAO survey, found that all tropical forest - wet and dry - was vanishing at a yearly rate of 7.3 million hectares (an area larger than Sierra Leone).

The forests usually go to make way for land-hungry farmers. But farming is often the worst use of TMF soils. Though lush, most hold nutrients in the vegetation, not in the soil. When cleared, most TMF soils are virtually sterile.

Governments often allow their TMF to be cleared and burned by settlers and cattle ranchers, logged out by timber companies and left bare by miners. They are selling a valuable resource cheap.

Intact TMFs are a warehouse of the genetic diversity needed to provide crop varieties to feed the world. They support known and as yet unknown medicinal, food and fuel plants. They protect watersheds, preventing floods, erosion and siltation.

As forests disappear, so do the tribal peoples who live in them. Brazil had 6-9 million Indians in the year 1500; it has barely 200,000 today.

A World Bank report suggests that only indigenous forest dwellers know how to derive a sustained living from TMFs. Outsiders, with their short-term profit goals, are usually doomed to failure.

In early 1982, UN agencies tried to draft a global "plan of action for the wise management of tropical forests". The attempt failed. Key TMF nations such as Brazil, Zaire, Colombia and Venezuela did not even take part.

> "Man has gone to the Moon but he does not know yet how
> to make a flame tree or a birdsong. Let us keep our
> dear countries free from irreversible mistakes which
> would lead us in the future to long for these same birds
> and trees." - President Houphouët-Boigny of Ivory
> Coast, a country in which more than 70% of the forest
> area in existence at the beginning of the century has
> been cleared.

Tropical moist forests are the world's <u>richest biological regions</u>, containing perhaps half of all types of living organism.

- But <u>almost half</u> of tropical moist forests (TMF) have already been <u>lost</u>, and the rest is being disrupted at a rate of 40 hectares (100 acres) a minute, according to a report by the US National Academy of Sciences.

- This irreparable loss is accompanied by the cultural, and in some cases the physical, extinction of scores of <u>forest dwelling peoples</u>. Their skills, social patterns and outlook constitute a fund of knowledge which industrialised man may need to call on in the future.

- TMF destruction results from:

  * <u>ignorance</u>: a failure to understand the high economic value of undisturbed TMF, and to realise that forest clearance is rarely the most efficient use of the land

  * <u>under-development</u>: a failure to provide growing populations with either enough jobs or agricultural land, the lack of which leads people to seek new and often unsuitable areas for farming

  * <u>helplessness</u>: few governments of TMF lands have the manpower, expertise or resources to effectively conserve their forests.

- TMF destruction:

  * deprives forest dwellers of their <u>traditional habitat</u>

  * could <u>disrupt fresh water</u> supplies for at least one billion people in three continents

* contributes to local and perhaps global <u>changes in climate</u>

* eliminates plants and animals which could provide mankind with future <u>crops and medicines</u>

* causes <u>flooding and siltation</u> of rivers and dams

* can lead to irreversible <u>loss of productive land</u>

* <u>reduces fuelwood</u> supplies and building materials for people who often have no alternatives

* is an uneconomic use of an irreplaceable resource.

- Tropical forests are <u>a valuable resource</u>. With growing populations needing land, the goods the forest can supply and the foreign exchange those goods can earn, TMF countries have pressing reasons for exploiting that resource. They resent it when developed countries, who have exploited their own forests for centuries, ask them to curtail their activities in TMFs on the grounds that tropical forests are a global heritage.

- But is large-scale deforestation the best use of the land on which the TMFs stand? And who benefits from current patterns of development?

- The problem is that forests are often seen only as a resource waiting to be exploited, whereas their <u>very existence has a value</u> which conventional economic models ignore. The enormous costs of deforestation do not become apparent until it is too late to reverse the process.

- Nations outside the tropics with little or no TMF land of their own are slowly realising that <u>they have a stake</u> in preserving these forests. In 1980, the then US president Jimmy Carter asked an inter-agency task force to prepare a US national strategy for dealing with global tropical deforestation.

- The task force's report listed the following reasons for US interest in the future of tropical forests:

* the global <u>scientific and environmental</u> value of TMFs, particularly regarding species loss and climate change

* US <u>investment</u> in and trade with TMF countries, as well as involvement in aid and development programmes in these countries

* US <u>consumer demand</u> for timber, medicines and other tropical forest products

* the <u>threat to internal US security</u> posed by social and political upheavals in TMF lands. (Deforestation in Haiti is a major cause of the recent influx of refugees to the US.)

CHAPTER 1

THE RESOURCE

Latin America contains 58% of the world's 900 million hectares (2.2 billion acres) of TMF, Africa 19%, Southeast Asia and Oceania 23%. A total of 70 countries contain TMF.

- Brazil alone possesses almost 33% of the world total, covering an area more than three times the size of France. Zaire and Indonesia each have 10%.

- Six other countries together contain more than 200,000 square kilometres (77,200 square miles) of TMF: Colombia, Peru, Venezuela, Gabon, Burma and Papua New Guinea.

- Two-thirds of the world total of TMF is rainforest. The rest is moist deciduous forest (containing mostly trees which shed their leaves annually).

- Rainforests are wetter (4,000-10,000mm (157-393 inches) of rain a year) than the deciduous group (1,000-4,000mm - 39-157 inches). They have more tree species which never shed their leaves, permanently moist soils, consistently higher temperatures (about 27 degrees centigrade - 81 Fahrenheit) and contain more diverse flora and fauna.

Three nations - Brazil, Zaire and Indonesia - possess 53% of the world's tropical moist forests.

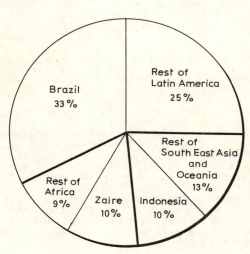

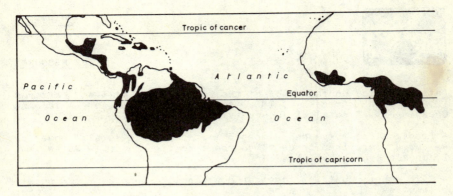

The planet's tropical moist forests (shaded areas) are yearly shrinking to make way for farms, roads, cattle ranches, logging schemes, mines and reservoirs.

- Inside untouched forests, there is perpetual twilight; the ground is covered only by nutrient-rich litter and shade-tolerant seedlings from the forest trees. There are occasional natural gaps in tree cover.

- Major concentrations of this primarily equatorial forest are in lowland Amazonia, the Congo basin, Sumatra and some Pacific islands.

- Deciduous forests contain fewer species but are more open to sunlight and thus have a denser undergrowth. They include monsoon forests in Burma, Thailand, Kampuchea, Java and Sulawesi, northeast Australia, parts of West Africa and South America; the subtropical rainforests of Central America and the Caribbean; coastal mangrove, high altitude ("cloud forests"), riverine and swamp forests.

- Forests can be divided into:

  * virgin: not necessarily untouched, but with a low level of human interference

  * secondary: growing up after primary forest has been felled. Within about 15 years, the total biomass equals that of the primary forest; then the vegetation gradually reverts to climax forest. This process may take hundreds of years. Patches of forest exist near the Angkor Wat temple in Kampuchea that were cleared 600 years ago and which are still different from the surrounding primary forest

  * degraded: when a forest area is interfered with to such an extent that the regenerative process cannot cope. In Indonesia, 20 million hectares (50 million acres) of once-forested land have been degraded to alang-alang grasslands through over-use.

Soil

The range of TMF soil types is extreme. Most TMF soil is poor, virtually sterile, but much of Java is covered by rich volcanic soil, and each year the floodplains of the Amazon (a tiny proportion of the Amazon TMF region) receive a new layer of fertile alluvial soil from the Andes.

- TMF forests have evolved effective mechanisms for recycling nutrients - from rain through leaves, the dense litter of the forest floor, humus (the organic constituent of the soil formed by the decomposition of organic matter) and the roof-mat. The roots of many TMF trees seek nutients by growing up out of the soil into the debris on the forest floor. This complex system makes possible the apparent paradox of lush forest growth on low fertility soils.

- Thus in most cases when a TMF area is cleared and burned, the nutrients are wasted; they go up in smoke, are blown away by the wind or run off into the streams and rivers. Worthless soil is left.

- Some soils with high iron and aluminium content, a common TMF condition, are liable to dry out and become brick-hard when cleared of trees and exposed to the sun. This process, sometimes called laterisation (the soils themselves are called "laterite"), is irreversible and often makes the exposed soil incapable of supporting vegetation.

- Clearing and exposure of the forest floor to the sun's heat inhibits the accumulation of humus. When the soil temperature exceeds 25 degrees centigrade (77 degrees Fahrenheit), humus decomposes faster than it forms, and the volatile ingredients, particularly nitrogen, are lost, robbing the forest floor of key nutrients.

- Even in the dry season in the humid tropics, 25mm (one inch) of rain can fall in 30 minutes - up to 40 times the amount in a typical temperate shower.

- In Ghana, four times more rain may fall in one hour in a typical storm (200mm - 7.9 inches) than London gets in an average month.

- A single storm can remove up to 185 tonnes of topsoil from a single hectare (75 tonnes/acre) which has lost its tree cover.

- But with TMF leaves acting as a shield, about three-quarters of the rain reaches the ground as a fine spray, and the rest trickles slowly down the trees, enriched by nutrients from the leaves and bark. Thus there is little danger of the sort of run-offs which cause erosion.

Plants and animals

The key role played by forests as genetic laboratories for medicine, agriculture and industry is generally unrealised and rarely taken into account when forest use is being planned or costed.

- Only about 1.6 million plant and animal species have been described and given scientific names; some scientists estimate that the worldwide total may be 5-10 million. Of this, 40-50% are believed to exist in TMFs. This means that up to half of the world's genetic diversity is concentrated on 6% of its land surface.

- On the most conservative figures, the TMFs are home to two million species, less than 1% of which have been scientifically examined for their potential value to Man.

- TMFs average 50-200 species per hectare (20-80 per acre); temperate forests are unlikely to contain more than 10 per hectare (four per acre).

  * Panama has as many plant species as the continent of Europe

  * Peninsular Malaysia has 7,900 species of flowering plants, compared with Britain's 1,430 species in twice the area.

- A survey of two hectares (five acres) of Brazilian Amazon forest found 1,986 plants over 1.5 metres (five feet) tall, representing 502 species. This is only about four individuals of any one species per hectare.

- Thus while the number of species is high, the density of any one TMF species may be very low, often little more than the critical level required to maintain a viable breeding population.

- Many plants, animals and birds are found in only one particular TMF area. So a small area of deforestation can wipe out entire species.

- An estimated 65% of the mammal species on the 4,500 square kilometre (1,700 sq mi) Indonesian island of Siberut, for example, are found only on this island, as are 15% of the plants.

- Indonesia, home of 17% of the world's bird species and 100 unique mammals, is one of 12 countries singled out for priority attention in the World Wildlife Fund's (WWF) 1982-83 tropical forests and primates campaign.

- Because of the stillness of the air under the forest canopy, few TMF plants are wind-pollinated. They rely on insects, birds and bats to spread their pollen.

- In many cases, a plant or tree depends on one pollinator species which in turn may be reliant on that plant for nourishment. The destruction of one species can assure the extinction of the other. Each of the 40 species of Central American fig trees relies on a different pollinator.

- Brazil nuts, of which the US imports more than $16 million worth a year, grow on trees that depend for the germination of their fruit on a rodent which chews and softens the seed coat. Brazil nut tree reserves must be large enough to support a breeding population of this rodent, or each seed coat must be softened artificially. Many experimental brazil nut plantations have failed because the trees are pollinated by one species of bee which requires other trees to feed on when the nut trees are not in flower.

- This complex system of interdependence is another reason why entire species of plants and animals can be threatened by relatively small areas of deforestation.

- There is a whole range of minimum critical sizes below which a forest reserve cannot support its constituent species. This size varies according to the feeding habits, breeding requirements, density and interrelationships of the species involved. Wide-ranging species such as certain mammals and birds need a larger space than most localised species.

- But a localised species may depend on a wide-ranging species for food or pollination and thus be unable to survive in a small reserve.

- Most TMF species, with their many such interdependencies and characteristic low densities (so that breeding populations are spread over a wide area), are likely to need a fairly large forest area for survival.

The people

In Latin America the Indians, who came to the continent 20,000-40,000 years ago, were the first - and until the 16th century, the only - inhabitants of the TMF. With the arrival of the Europeans, the rainforest Indians retreated deeper into the forest. Many remained in isolation for 400 years and some are still isolated.

- The aborigines of the African TMF were <u>Pygmies and Khoi-Sah</u> (Bushmen and Hottentots). Several hundred thousand Pygmies still live in the Central African TMF, but the Bushmen now live only in the Kalahari Desert.

- Latin American forest dwellers moved into the forests from a more <u>agricultural way of life</u>. Both Amerindians and Pygmies now practise agriculture as well as hunting and gathering.

- In <u>Southeast Asia and Oceania</u>, there has been such a cross-fertilisation of populations among and between islands and the mainlands that it is impossible to say who were the original inhabitants of any area. Many groups have been living in the forest for thousands of years, and agricultural and hunting techniques have been modified as groups have intermingled and intermarried.

- The exact number of forest dwelling groups is not known. Their overall numbers have been <u>drastically reduced</u> in the past century as outsiders have moved in to "develop" their ancestral lands, which has meant the destruction of the aborigines' environment.

- Brazil's Indian population was estimated at 6-9 million in 1500. It had dropped to one million by 1900 and to under 200,000 today - an attrition rate of <u>two million people per century</u>, according to the World Bank.

- Despite the devastation, <u>many forest tribes remain</u> - a few probably unknown to the rest of the world, as were the Tasaday tribe of the Philippines until 1972.

- During the building of Brazil's Transamazonian Highway, previously unknown groups of forest dwellers were encountered at a rate of <u>one per year</u>.

- Of the 230 tribes living in Brazil in 1900, about 143 survive. Colombia has about 180 language groups. New Guinea has 800 tribes, mostly forest dwellers, each with its own language. In fact, <u>one fourth of the languages</u> of mankind are found in New Guinea.

- Forest tribes are distinguished by their <u>special relationship</u> with the forests. Their culture has evolved in harmony with the forest environment, and their identity is tied up with life in the forest.

- Traditional forest dwellers who hunt, fish, grow crops, gather food and use trees to make homes and canoes <u>do not degrade the forest</u> unless their population density becomes greater than the forest can support.

- Overpopulation is inhibited by <u>cultural mechanisms</u> which may include contraception, sexual taboos, abortion, infanticide and warfare. (It is also inhibited by disease.)

- On some rainforest soils, the forest dwellers are the only people to have evolved sustainable <u>long-term productive systems</u>.

- A World Bank report on "Economic Development and Tribal Peoples" published in July 1981 refers to "the great potential value of tribal knowledge of management of marginal lands". This knowledge, it says, represents "an <u>increasing investment opportunity</u> contributing significantly to the dominant society".

- Not realising the <u>diversity and complexity</u> of the cultures developed by forest tribes, most outsiders see these tribespeople as beings who have no place in modern society.

- In fact, forest dwellers' cultures, like those of peoples elsewhere, evolve and change in response to circumstances. The question is whether they will be allowed to <u>engineer their own adaptations</u> to the modern world or - if they are allowed to survive at all - be forced to abandon their cultures by the dominant society's insistence on uniformity.

An 1935 print of Brazilian TMF being cleared for a coffee plantation. Despite steady deforestation over the years, the threat of serious global TMF depletion has emerged only in the past decade.

Photo: Marcos Santilli/Earthscan

CHAPTER TWO

THE PROBLEM: DEFORESTATION

"Forest formations have been considered as an endless reserve of land, fruits, animals and wood. Recent history shows us that these resources, theoretically renewable, can also vanish if care is not taken to avoid using them beyond critical thresholds.... In the long run, there is a real ecological threat to the whole (of Africa), and mankind should be concerned about what will be left of the forest cover of Africa in the second half of the twenty-first century."

- This was the verdict on Africa of an FAO report, "Tropical Forest Resources Assessment Project 1981". But the warning applies to all the world's tropical moist forests.

- Estimates of TMF loss vary considerably, but over the past few years estimates have come closer together.

- By the mid-1970s, TMF covered 935 million hectares (2.3 billion acres), and had been reduced from their natural global coverage by about 40%, according to a 1977 FAO report by Adrian Sommer.

- Left alone for several centuries, some of the deforested area could revert to its original form. But the pressures that led to its exploitation are unlikely to allow a long period of rehabilitation. Overuse leading to severe degradation is more likely to be the fate of cleared or depleted TMF.

- Studies by the World Bank and the Commonwealth Forestry Institute have indicated that over 1963-73 about 150 million hectares (370 million acres) of TMF were lost, or 14% of the pre-1963 total. A 1976 FAO report had put the deforestation rate at 11 million hectares (27 million acres) per year.

- Thus from a state of relative equilibrium, TMFs are suddenly under threat.

- The most optimistic estimate of annual loss is 5.6 million hectares (13.8 million acres), by Lanly and Clement in 1979, but this includes only the permanent and complete elimination of forests.

- At the other end of the scale, "Conversion of Tropical Moist Forests", published by the US National Academy of Sciences in 1980, said that over 20 million hectares (50 million acres) were being destroyed or seriously degraded every year. These figures include selective logging projects.

- At that rate the most accessible areas would be converted to other land use or to wasteland by the year 2000.

- The FAO, under the auspices of UNEP's Global Environmental Monitoring System (GEMS) carried out a survey in 1981 which concluded that tropical deforestation (which includes other types of tropical forest besides TMF) in the late 1970s totalled 7.3 million hectares (18 million acres) per year, or 14 hectares (35 acres) per minute.

- This consisted of an annual decline of 4.2 million hectares (10.4 million acres) in Latin America, 1.8 million hectares (4.4 million acres) in Asia, and 1.3 million hectares (3.2 million acres) in Africa.

- The rate of deforestation varies from country to country, region to region. Some TMFs remain virtually undisturbed, while others face extinction.

- Recent satellite photos from Brazil's Amazonian TMF show, to many experts' surprise, that only 2% of the area had been cleared as of the mid-1970s. But these photos also show extensive, ill-planned clearance along the southern fringes of the rainforest, along roads and around towns, and in a large eastern zone, according to Erik Eckholm in "Down to Earth", (W W Norton & Co, 1982, New York, and Pluto Press, London).

- Also, photos from recent years show that clearance of this 280 million hectare (690 million acre) area is running well above the one million hectares (2.5 million acres) per year deforestation which had been commonly assumed. Also, relative to their forest areas, countries such as Peru, Colombia, Ecuador and Venezuela - occupying the upper tributaries of the Amazon system - are losing forests faster than Brazil.

- African deforestation ranges from an insignificant 0.2% a year in Zaire to 10% in Nigeria and Ivory Coast.

- Deforestation in Africa (in the sense of clearance for agriculture and other land uses) is most serious along the west coast from Guinea-Bissau to Nigeria, a region in which only 18 million hectares (45 million acres) remain, with an average annual deforestation rate of 4%.

- Overexploitation for fuelwood, overgrazing and repeated fires take a further toll, and tree plantations successfully established every year represent only 2.5% of the forested areas cleared annually.

- But faced with modern pressure for land and timber, no TMFs are safe.

- Asian seasonal forests have been exploited most intensively, because of population pressure and because the commercial value and homogeneity of the continent's dipterocarp species makes logging attractive. (The Dipterocarpaceae family, so called because its fruit has two "wings", have tall straight trunks, few branches and produce valuable hardwoods.)

- But with many Asian forests on the verge of commercial extinction after 20 years' intensive logging, timber companies are turning their attention to the less well-known Latin American species. The journal "Import/Export Wood Purchasing News" has described it as a "veritable race to Latin America and in particular to the Amazon Valley".

- The pace of change which can occur is illustrated by the Indonesian experience. Log production there multiplied <u>more than sixfold</u> between the period 1961-65 (4.1 million cubic metres - 145 million cubic feet) and 1976-79 (25.9 million cubic metres - 915 million cubic feet). In the same period, log exports grew from 125,000 cubic metres (4.4 million cubic feet) to 19 million cubic metres (671 million cubic feet). Domestic processing increased from 5,000 cubic metres (177,000 cubic feet) in 1968 to 526,000 cubic metres (18.6 million cubic feet) a decade later.

- Recent satellite pictures of the Philippines, traditionally a major timber exporter, show that forests now <u>cover only 30%</u> of the country, though the government feels that <u>46%</u> coverage is a minimum for both economic and environmental reasons. A consortium of Philippine research groups has warned that the nation may not be able to produce enough timber for even domestic consumption by the year 2000.

CHAPTER THREE

THE EFFECTS

By conducting a steady and <u>moderate flow of rainwater</u> to major rivers and lakes, and by <u>giving protection with their roots</u> and foliage against flooding and landslides, TMFs contribute to the health and safety of hundreds of millions of people.

- The danger of erosion after the removal of forest cover is usually <u>greater in TMFs</u> than in temperate regions, as rainfall is more intense and the soil base often shallower. Compaction by rainfall after clearing decreases the absorptive capacity of the soil, further increasing the danger of flooding and landslides.

- In the Amazon, an average annual rainfall of 2,160 mm (85 inches) removed under half a tonne of soil from every 4,000 square metres (43,000 square feet) of a forested slope of 12-15 degrees over three years. When an area with only half the incline <u>was cleared</u>, the run-off rose to 45 tonnes.

- Deforestation in northern Luzon, Philippines, has silted up the Ambuklao dam reservoir, <u>halving its expected life</u> from 60 to 32 years.

- A man-made irrigation lake, Meiktila, in Burma, functioned unchanged and undamaged for over 800 years, until this century when <u>traditional taboos</u> against clearing forest within 3.2 kilometres (two miles) of rivers were ignored. Now the lake has begun to silt up: between 1926 and 1951 the area irrigated fell by half to 11,000 hectares (27,200 acres).

- In Colombia and Costa Rica, deforestation has resulted in siltation and reduced capacity in <u>hydroelectric reservoirs</u> and in electricity rationing.

- Deforestation in the Panama Canal Zone has caused siltation that has reduced the storage capacity of the lakes and reservoirs <u>on which the canal depends</u>. Lakes Gatun and Madden not only feed the canal, they also provide several urban centres in Panama with hydroelectric power and domestic water supplies.

- Deforestation in Panama has also resulted in an alternating pattern of <u>floods and droughts</u> that has made water supplies unreliable.

- Parts of Thailand's traditional waterway transport system have been clogged by siltation. The government is so alarmed by the consequences of deforestation that it has threatened severe punishment, including summary execution, for illegal felling of trees. Every year about 30 Thai <u>forest guards are killed</u> in gun battles with hardwood poachers.

- Studies in Amazonia have indicated that streams whose banks are not lined with shrubs and trees show an increase in <u>turbidity, nutrients and heat</u>, conditions unfavourable to fish, Amazonia's chief source of protein.

- The Amazon basin supports more fish species than any other river system. If the myriad rivers and streams feeding the Amazon cannot support this fish life, <u>thousands of species</u>, many with unexploited commercial and nutritional potential, may disappear.

Local climate

The links between <u>TMFs and climate</u> are thought to be close, but are not well understood.

- Up to half the rain in the Amazon basin is generated by water evaporating <u>from the forest</u>, not from winds coming from the ocean.

- Forest destruction in Haiti and Senegal is believed to be linked to significant <u>decreases in rainfall</u> over the cleared area.

- After the removal of forest canopy, more light and heat are reflected back into the atmosphere in what is known as the "<u>albedo effect</u>". With the forest no longer shading the ground by day and releasing heat at night, surface temperatures become more extreme - hotter by day, colder by night.

- These interactions will affect and be affected by other consequences of deforestation, such as decreased soil absorption and extremes of flooding and drought, in a <u>series of chain reactions</u> that are hard to predict and impossible to prevent.

18.

Global climate

Scientists <u>do not agree</u> on the effects of widespread deforestation on the global climate.

- It is a widespread myth that rainforests, with their masses of living vegetation converting carbon dioxide to oxygen, produce a large proportion of the Earth's oxygen. In fact most scientists agree that mature TMFs do not contribute to the global supply of oxygen; they are in a <u>state of equilibrium</u>, consuming through the decay of organic matter as much oxygen as they produce through photosynthesis.

- The effect of deforestation on the <u>global carbon budget</u> is more contentious.

- Cutting and burning forests, which are reservoirs of stored carbon, add to the concentration of carbon dioxide <u>in the atmosphere</u>.

- Atmospheric carbon dioxide has increased by about 15% in the last century, one-third of this since 1960. The burning of <u>fossil fuels</u> has been a major contributor, but widespread forest clearance adds to the increase.

- A buildup of carbon dioxide in the atmosphere could trap solar radiation by what is called the "<u>greenhouse effect</u>", causing a rise in temperature (though not necessarily uniformly around the world).

- The often-predicted doubling of atmospheric carbon dioxide by the middle of the next century might raise the global temperature about <u>two degrees centigrade</u> (four degrees Fahrenheit), which would <u>make the Earth warmer</u> than it has been for the past 1,000 years.

- There is a <u>strong scientitific consensus</u> on this figure of two degrees centigrade. There is less agreement on the possible effects of this temperature increase.

- The warming would probably be weighted <u>towards the poles</u>, so that conditions favourable to temperate agriculture could shift from its present latitudes, a development which might cause Siberia to replace the US Midwest as the world's major grain producing region.

- Temperature changes could affect <u>rainfall patterns</u>, causing less rain in some major agricultural areas.

- A temperature increase of five degrees centigrade (nine degrees Fahrenheit) might partially melt the West Antarctic ice cap, with the consequent rise in sea level perhaps inundating <u>the world's coastal areas</u>, where many large cities are located.

- A <u>feedback effect</u> might occur if, as a result of warming, the oceans released additional carbon dioxide into the atmosphere. This process could continue indefinitely, and is believed by some scientists to have been responsible for a temperature rise between the Mesozoic period (which began 225 million years ago) and the Cainozoic (70 million years ago), when three-quarters of the Earth's species disappeared.

- This scenario of disasters is by no means accepted by all of the scientific community. The effects of increased atmospheric carbon dioxide <u>cannot be predicted</u> at present, and the extent to which deforestation is contributing to the increased concentration is not known.

- But as George Woodwell, director of the Ecosystems Center of the Woods Hole Marine Biology Laboratory in Massachusetts, USA, has pointed out, by the time we have conclusive proof it will be <u>20 years too late</u> to take appropriate action.

<u>Peoples of the forest</u>

As the pressure on the forests increases, so does the pressure on the <u>traditional forest dwellers</u>.

- The disastrous experiences of most newly-contacted tribes have reinforced the notion that they are doomed in any confrontation with the developed world. Many groups have been overwhelmed by the <u>diseases, weapons and technology</u> of the dominant culture.

- Demoralisation accumulates as a tribe loses faith in <u>its religion and medicine</u>, learns to depend on goods which it cannot produce, and realises that even its leaders cannot negotiate effectively with outsiders who hold them in contempt.

- Anthropologists or officials who try to help forest-dwellers by giving them food, medicine or tools often reinforce their self-image as helpless children - forest tribes in many countries are legally <u>classified as minors</u> - thus furthering the break-up of once independent cultures.

- The process <u>is self-perpetuating</u>: loss of traditional skills makes it difficult to make a living from the forest; malnutrition causes greater dependency on imported goods; moral dependency on the values of the dominant culture increases. As traditional values are forgotten, a tribe's ability to exist as an independent unit is further reduced.

- Most governments regard the traditional forest groups as a nuisance. Where there is an official policy, it is usually one of "<u>integration</u>" into the dominant culture.

- The then Brazilian Interior Minister, Mauricio Rangel Reis, said in 1974, "We are going to create a policy of integrating the Indian population into Brazilian society <u>as quickly as possible</u>.... We think that the ideals of preserving the Indian population within its own habitat are very beautiful ideas, but unrealistic."

- Officials tend to see the integration of cultural minorities as a way of increasing <u>bureaucratic efficiency</u>. Administration is easier if all groups conform to the same norms. Nomads are difficult to organise, administer and tax. Collective land ownership is a foreign concept to most governments.

- Traditional groups are widely believed to occupy more than their fair share of land. This evokes resentment from both small settlers and large commercial concerns.

- Some of the worst abuses have been suffered by the Latin American Indians. Desire for land, coupled with exaggerated fears of Indian fierceness and a widespread belief that they are subhuman, have led to many atrocities.

- A 1968 Brazilian government investigation of the subsequently disbanded Indian Protection Service found that Cintas Largas (Surui) Indians in the northern Mato Grosso had been bombed from the air with dynamite, and most of the Beicos-de-Pau tribe died after its food had been laced with arsenic.

- Tribes in Mato Grosso had been the victims of germ warfare between 1957 and 1963 when tuberculosis, influenza, and smallpox were deliberately introduced into their villages. The Arara tribe fell from 152 members to 34 after accepting poisoned sweets from jaguar hunters, the investigation found.

- A reliable European development aid agency recently cited evidence from anthropologist Anna Presland that the Arara were bombed by army helicopters in 1979, and that the army used bombs and machine-guns against the Waimiri-Atroari group in opening land for the BR-174 section of the Transamazonian highway in the early 1970s.

- Dr Mark Munzel, an anthropologist at Geissen University, West Germany, charged in studies published in 1972 by the International Working Group for Indigenous Affairs (IWGIA) that the Ache (Guayaki) Indians of eastern Paraguay near the Brazilian border were the victims of systematic genocide. One of the last groups of hunter-gatherers in South America, they need large areas of game-rich land to survive.

- Dr Munzel, and later other witnesses, said that until the late 1970s the Ache were the objects of manhunts in which the Paraguayan army sometimes participated. Professor Miguel Chase Sardi, an anthropologist at the Catholic University, Asuncion, charged that during the hunts the parents were killed and the children later sold.

- The survivors not sold into slavery were placed in the Ache reservation, run since 1972 by the Protestant New Tribes Mission in cooperation with the Paraguayan government. Here the Ache are discouraged from following their traditional way of life.

- The Ache manhunts have apparently ended - largely because there are no more "free" Ache - but there is still evidence of Aches being used as slave labour.

- Survival International, a London-based group which champions the causes of the world's tribal peoples, recently charged that

another Indian tribe, the Ayoreo, has been the object of <u>further</u>
<u>manhunts</u> by New Tribes Missionaries at the El Faro Moro <u>mission</u>
in northern Paraguay.

- It said that Indians were spotted in the forest by aircraft,
and what the missionaries call <u>"tame" Indians</u> (that is, converted
to Christianity) were sent in with guns to bring the forest
Indians back to the mission. There, many died from shock,
inadequate food and lack of medical care.

- The leader of one rounded-up group died within a few weeks.
Missionaries explained to an outsider that <u>"he wouldn't eat</u> the
food we gave him and we don't have a whole <u>lot of jungle</u> food
lying around".

- The 3,200 Yagua Indians who live along the borders of Brazil,
Colombia and Peru have suffered in a different way. They have
been forced to rely on "patrones", middlemen, who buy their
services or goods cheaply, keeping them in a <u>permanent debt</u>
<u>bondage</u>, and encouraging them to abandon their traditional
communal houses in favour of a central settlement over which it
is easier to establish control.

- In 1976 contractors brought a <u>whooping cough epidemic</u> to one of
the last communal groups of Yagua. So many died that the rest,
who were forced to abandon the settlement, simply moved into
shacks, too demoralised to start a new community.

- Since 1962, <u>tourist agencies</u> in Iquitos, Peru, and Leticia,
Colombia, have offered excursions to view the "exotic" Yagua.
Tourists supply the Indians with liquor and then take snapshots
of their drunken behaviour.

- Governments occasionally intervene to stop harassment or
persecution of forest tribes when it is carried out by tourists,
poachers, small-scale prospectors or farmers. But the more
serious disruptions are often <u>with government blessing</u> and
assistance, and in the name of progress and development.

- In March 1974, President Ferdinand Marcos of the Philippines
issued a decree <u>reserving "ancestral lands"</u> to the nation's
cultural minorities. Missing from the list of minorities was the
Tinggian tribe of Abra.

- Ten days later the Cellulose Processing Corporation was awarded
100,000 hectares (247,000 acres) of forest land, much of it the
<u>homeland of the Tinggians.</u> They have subsequently tried
unsuccessfully to be included on the official list of cultural
minorities.

- Relations between the Tinggians and the government-owned parent
company, Cellophil, are bad. Replanting of logged-out areas has
been slow, effluent from a pulp mill has <u>polluted the Abra River</u>,
and the floating of logs down rivers has destroyed irrigation
canals, dykes and crops.

- Many Tinggians have been detained for their refusal to accept
Cellophil's presence on their land, and <u>four Catholic priests</u>
have left their parishes to join the rebels.

- The Philippine government is also planning a series of dams along the Chico and Agno rivers to attract more industry to the region.

- The homes and rice fields of 100,000 people will be flooded by the first scheme - the $50 million Chico River dam in northern Luzon - as will their sacred ancestral graves. The families will lose their land, and thus the social, economic and religious basis of their way of life.

- The side-effects of the dams and the export-oriented industries they will attract are predictable: pollution of the rivers and loss of fish, and deforestation resulting in erosion, flooding and unstable irrigation for rice fields.

- Opposition has been fierce, with widespread fighting between tribesmen and government soldiers. Roman Catholic Bishop Francisco Claver said in a letter to President Marcos: "The people seem to have a far richer and more comprehensive notion of human development than that espoused in practice by the government.... What the government defines as development for them is actually the destruction of all that they value in life, above all their spirit as a people. This spirit is no small thing for survival itself."

- In an open letter to Robert McNamara, then president of the World Bank, tribal elders accused the Presidential Assistant on National Minorities (PANAMIN) of using force, intimidation and bribery to induce villagers to acquiesce in the project.

- Land is at the root of many of these problems. The forest homeland is not only essential for a group's physical survival, but also provides a secure spiritual and emotional base from which forest dwellers can organise their own response to outside pressures.

- Deprived of their land, forest dwellers can only look forward to absorption into the lowest level of the dominant culture as beggers, landless peasants or low-paid labourers, like the Conibo Indians in Peru who have been reduced to working for the logging company which gained control of their ancestral lands.

- The London-based Anti-Slavery Society reported in 1977 that the Chiquitano Indians of Bolivia were being induced with promises of food and clothing to work on rubber plantations. There they quickly fell into debt at company stores where prices were five times higher than normal.

- Debts were considered by the management to be passed on from father to son. At night the workers were locked up in sheds, and sometimes handcuffed. In 1976, a group of 25 Indians escaped from one such camp, only to be returned to the owner by the police. A government report acknowledged the existence of this bondage system.

- Land laws vary enormously:

   * in Papua New Guinea 97% of the land is held under customary ownership by clans or villages. Villagers can be put under

pressure to acquiesce in a project, but some tribes have blocked developments they find objectionable

* several <u>Latin American governments</u>, including Brazil and Guyana, have committed themselves to assigning forest lands to their traditional owners, but have failed to carry out their promises

* <u>Peru</u> claims to guarantee the integrity of the land of native forest-dwelling communities, but it also provides for free access for oil and gas pipelines and mineral exploitation, with no compensation to the occupiers of the land

* neither <u>Ecuador nor Bolivia</u> have any special provisions for recognition of ownership of traditional Indian lands, nor have they set aside any reserves for Indian occupation

* while individual members of a tribe may establish title to a plot of land under <u>Bolivia's Agrarian Law Reform</u>, there is no mechanism for assigning collective ownership, a notion crucial to most Indian cultures.

- The Yanomamo, one of the last large Latin American Indian tribes <u>living relatively undisturbed</u> (10,000 in Brazil and an equal number in Venezuela), have suffered greatly from the invasion of their lands by roads, squatters, miners and prospectors.

- Deposits of uranium and cassiterite (the raw material for tin) were found on Yanomamo land, leading in 1975-76 to an influx of miners and, according to local missionaries, "<u>the murder of Indians</u> who insisted on remaining in the region of the mining site".

- The government eventually ordered the individual miners out, but has allowed <u>a number of large companies</u> to conduct surveys in heavily populated Yanomamo areas, including a region containing some 6,000 Indians who had virtually no previous experience of contact with outsiders and had not been vaccinated against the diseases often unwittingly brought in by outsiders.

- In Brazil, spokesmen for the Yanomamo Indian Nation have presented a formal complaint to the Inter-American Commission on Human Rights of the Organisation of American States, <u>charging the government</u> with disregarding basic principles of human rights in its treatment of the Yanomamo. The government was also accused of violating the constitution and several international conventions by failing to assign to the Yanomamo the lands they have inhabited for centuries.

- In 1982 the government established a Yanomamo Park, but the Indians note that often such parks exist <u>only on paper</u>, and reserves have often been abused by governments.

- The World Bank has recently loaned Brazil money for a development project which includes a road through seven Nambiquara Indian villages and development of an area in which some 8,000 Indians live. The 530 inhabitants of the threatened villages are the <u>sole survivors</u> of the 10,000-15,000 Nambiquara who traditionally <u>hunted and farmed</u> on those lands.

- Yet the World Bank report "Economic Development and Tribal People" noted how during the 1970s roads brought more than 20 government-promoted agribusiness ventures into the Nambiquara villages of the Guapore valley, and "diseases decimated the Nambiquara to the point that in two of the Guapore valley groups the entire population younger than 15 years was <u>killed by influenza and measles</u>".

- Since their first large-scale contacts with outsiders at the turn of the century, the history of the Nambiquara has been one of <u>violence, deception and disease</u> brought to them by people who wanted their land. Most of the Nambiquara's lands have been assigned to large cattle ranchers who have sprayed herbicides to clear it, making it useless to the Indians.

CHAPTER FOUR

THE CAUSES

One underlying cause of deforestation is <u>rapid population growth</u> in TMF countries.

- People need wood <u>for fuel and building</u>. As an FAO report on tropical forests noted, "in the absence of organised supply, rural populations resort to indiscriminate firewood collection, excessive lopping of trees and overgrazing, which lead to depleted resources, sometimes beyond repair".

- For countries with overwhelmingly rural populations - which are expected on average to double within 27 years - the most obvious resource of a forest is <u>the land the trees stand on</u>.

- The concentration of most of the agricultural land in the <u>hands of a few people</u> also increases pressures to open new areas.

  * Some 93% of Latin America's arable land is concentrated in the hands of <u>7% of the landowners</u>, and millions of farming families possess no land at all.

  * In Brazil, 70% of the 120 million population are <u>entirely landless</u> or lack secure access to land. The figure for Java is 85%.

  * In the <u>Brazilian Amazon</u>, two kinds of landholding have increased <u>dramatically</u> since 1960: those of over 10,000

hectares (25,000 acres) and those of under 10 hectares (25 acres).

* In El Salvador, with a population of 4.5 million, until recently more than 40% of the land was in the hands of under 200 families.

- New sources of unoccupied land - unoccupied except by the forest dwellers whose rights and existence are often hardly acknowledged - is a safety valve welcomed by governments, the rich and the despairing poor.

- But often the new land proves worthless for the purposes to which the settlers want to put it. And in their desperation for quick solutions to the problem of overcrowding - solutions which avoid sweeping political or social changes - governments have allowed and even encouraged settlers to move in without the planning and back-up which might have made successes of some of the colonisation failures.

Conversion to agriculture

Clearing land for farming is the main cause of forest loss today.

- Traditional shifting cultivation involves clearing a small area of forest, burning some of the felled vegetation and leaving the rest to decompose and gradually release nutrients to the soil. By growing a mixture of crops, farmers can make maximum use of nutrients. Nevertheless, productivity on most TMF soils declines markedly within two or three years, and the farmer moves on to clear a new patch.

- As long as population density remains at two or three people per square kilometre (five people per sq mi) and land can lie fallow for at least 10 years, shifting cultivation is a sustainable and self-contained system.

- After letting a forest area lie fallow for, say, 10 years, farmers return to the area and clear the secondary forest which has sprung up in the meantime. This system takes the pressure off the primary forest. Many forest dwellers find that secondary forest is easier to clear and contains more suitable materials for construction and food.

- Often the secondary forest is seeded with fruit or timber trees, so is useful even in fallow. Some animals, including elephants and gorillas, prefer the mosaic of young vegetation and mature forest created by shifting cultivation. The young vegetation is a productive storage ground; the mature forest provides shelter, and together they support much higher densities of animals than does undisturbed forest.

- These animals provide forest dwellers with a convenient protein pool which enables them to consume as much animal protein as do citizens of industrialised countries.

A TMF area burned for farming in the Rondonia area of western Brazil. Here an influx of settlers has turned once lush forests into a wasteland of erosion and brick-hard soil. (see page 27)

Photo: Marcos Santilli/Earthscan

- Many forest cultures that rely on shifting cultivation have developed sophisticated techniques for maximising agricultural productivity. Forest cultivation has also frequently been found to require <u>less labour per unit of return</u> than conventional agriculture, thus leaving time for supplementary hunting, fishing and gathering.

- Problems arise when newcomers raise the population:land ratio to unworkable levels, practise shifting cultivation <u>without rotation</u>, or lack the expertise of the traditional forest dwellers.

- Instead of leaving a sufficient fallow time and using other techniques that help the land retain its productivity, new settlers often exhaust the land to the point at which it is <u>permanently degraded</u> and useless for agriculture or forestry. They then move deeper into the primary forest.

- In TMF countries with <u>low population densities</u>, such as the Central African Republic, Gabon and Equatorial Guinea, agricultural clearing accounts for less than 0.1% of their total area of closed forest each year.

- In other countries, however, conversion to agriculture is a
major problem. UN specialists describe the position of Nigeria,
Ivory Coast, Burundi, Rwanda and Madagascar as "alarming".

- For Nigeria and the Ivory Coast, where the deforestation rate
is about 10% a year, clearing for agriculture "has significantly
jeopardised the future of forestry in these countries and wasted
a considerable potential wealth, much higher than that of the
logs extracted before clearing" (Forest Resources of Tropical
Africa, FAO, 1981).

- West African farmers clear 1,900 hectares (4,700 acres) of
productive closed forest every day by axe and fire. (Closed
forests are forests in which tree crowns cover at least 20% of
the ground when viewed from above.)

- In Latin American TMFs, shifting cultivation with rotation
accounts for only 33% of annual deforestation. Thus the
remaining 67% of the area deforested annually is stripped of
trees for more or less permanent use and will not be allowed to
revert to forest.

- Many areas in Latin America have been deforested through
government-sponsored colonisation schemes, as most governments
regard forest land as a reserve of "unused" land suitable for
development.

- The effects of unwise clearing can still be seen in two sites
in Brazil, which were settled in the first half of this century.

   * In a successful effort to encourage colonisation, the
     Brazilian government laid a railway in the forest-rich
     Bragantina zone of the Amazon in 1883-1908. By the late
     1940s, the zone had one of the highest population densities
     in Amazonia, and 30,000 square kilometres (11,580 sq mi) of
     forest had been converted to agricultural land. But the
     land could not support intensive cultivation, and today the
     Bragantina is a vast unproductive expanse of scrubland and
     laterite, a semi-desert that has been described as a "ghost
     landscape".

   * When it became clear that the Bragantina could not provide
     for a large population, the government relocated some of the
     peasant colonisers to the Amazonian territory of Rondonia.
     After three years of relatively good agricultural returns,
     the soil started to show signs of exhaustion; erosion and
     laterisation began.

- Indonesia is engaged in a "transmigration" scheme which seems
to be repeating many of the Brazilian mistakes of a decade ago,
although ecological studies and planning are now being taken more
seriously.

- Two-thirds of Indonesia's 149 million people are crammed onto
the island of Java, whose rich volcanic soils support the most
intensive agriculture in the world. The average population
density of 600 people on each square kilometre (1,500 people on
each sq mi) rises to well over 1,000 per square kilometre (2,600
per sq mi) in irrigated areas.

- Half of Java's rural population is landless; another 20% has insufficient land to provide a subsistence living. A combination of spontaneous and official colonisation has been populating some of Indonesia's other 13,000 islands since the turn of the century. Now a campaign is underway to move 500,000 families to 250 sites on less crowded islands by 1985.

- Although the target numbers would not even keep pace with Java's annual population increase of over two million, the programme moved only 60,000 families between 1974 and 1979.

- The programme, partly funded by the World Bank, will cost several billion dollars ($537 million in 1981-82 alone). Although the necessary soil, weather and ecological studies have not been carried out, existing information indicates that most of the outer islands' land capable of supporting permanent agriculture has already been settled.

- Most settlement so far has been in forested areas. But a 1979 presidential decree prohibited clearing of primary forest for migration, so the goal is now to concentrate colonies in areas covered with imperata grass (alang alang). There is no sound technology for creating a sustainable agriculture on such lands, and many settlers will probably be forced to return to shifting cultivation in forests to feed their families.

- The project has been plagued by conflicts among natives, settlers who have moved on their own initiative and government-sponsored colonists receiving aid and technical assistance. Few of these colonists are experienced farmers, since the scheme takes two-thirds of its participants from the poorest rural dwellers and another 10% from the urban homeless. Religious, tribal and linguistic differences also hamper the project. The quality and quantity of government support for the settlers was questioned by the International Labour Organisation.

Timber

After shifting cultivation, timber harvesting is the second main contributor to deforestation in the TMFs of Asia, Oceania and Africa. It is on the increase in Latin America. In fact, the cultivators often follow the timber companies into the forests.

- The FAO estimates that one million square kilometres (390,000 sq mi) of tropical forest was leased for commercial timber extraction between 1958 and 1978. The three main uses for cut trees are building materials, pulp and fuel.

- Fuelwood cutting is a relatively minor problem in TMF conversion. But as dry woodlands and savannahs are exhausted and population in TMF regions increases, pressure on these forests will grow unless there is a rapid expansion of village woodlots for fuel.

- The World Bank estimates that the rate of planting for fuel needs alone must <u>rise five-fold</u> from the present 500,000 hectares (1.2 million acres) a year in the Third World (excluding China) if major ecological and economic costs are to be avoided. But this planting must be done not in forests but in the villages and on the farmlands where people live.

- About 80% of forest production in developing countries (47% worldwide) <u>is used as fuel</u>, and an estimated 150 million cubic metres (5.3 billion cubic feet) of fuelwood is taken from TMFs each year. But it is impossible to know to what extent this contributes to TMF deforestation since much depends on the concentration of fuel gathering, the size of harvest which an area of forest can sustain and the proportion of fuelwood supplied by planted trees.

- In Brazil some four million tonnes of charcoal are used annually in the <u>steel industry</u>, and the intention is to extend such consumption, reducing dependence on imported coke and coal, according to the 1980-81 FAO Commodity Review and Outlook.

- As much as 500 million cubic metres (17.6 billion cubic feet) per year of usable wood <u>is burnt on the spot</u> after clearing to save transport and storage problems and to provide soil nutrients (US National Academy of Sciences, 1980).

- Just over 10% of the world's supply of industrial wood (timber and pulp) comes from TMFs. Tropical hardwoods are increasingly <u>replacing temperate hardwoods</u> in the building and paper industries as supplies of the latter dwindle and environmental safeguards for temperate forests are strengthened.

- Since 1950, when exploitation of tropical hardwoods began to take off, imports by developed countries have <u>increased 16-fold</u>.

- <u>Roughly half</u> of all tropical hardwood timber is exported - 53% to Japan, 32% to Europe and 15% to the US.

- The FAO predicts that by the year 2000, production of tropical hardwoods will have <u>more than doubled</u>, with developed countries requiring two-fifths of total output.

- More than 70% of all tropical hardwoods are produced by <u>six countries</u>: Indonesia, Malaysia, Philippines, Papua New Guinea, Brazil and Ivory Coast. A further eight countries bring the total to 90%: Colombia, Ecuador, Gabon, Ghana, Nigeria, Costa Rica, Burma and Thailand.

- The Centre for Agricultural Strategy at Britain's Reading University projects that tropical hardwood demand <u>will exceed supply</u> by 400 million cubic metres (14.1 billion cubic feet) in the year 2000 and by three billion cubic metres (106 billion cubic feet) 25 years later.

- Demand for <u>paper products</u> is expanding more rapidly than any other industrial wood use.

- TMFs currently produce 7% of the world supply of paper pulp. That figure is likely to rise dramatically as the developed countries increase their already high demands (now largely met internally), and developing countries' populations and literacy rates grow.

- TMFs, with their relatively low proportion of softwoods, were previously not considered suitable pulp producers. But new techniques have made it possible to combine wood chips from 100-200 TMF hardwood species to produce pulp.

- Asia is the main timber and pulp producing region, with 67% of total output, and the chief exporter, with 86% of all exports.

- Virtually all Indonesia's easily accessible lowland forest, including some areas designated as "protected forest", have been let as timber concessions.

- Most companies like to finish their extractions in 4-10 years rather than establishing long-term operations in countries where they consider the political and economic future to be unpredictable.

- Government-imposed rules limiting felling or requiring reforestation by timber companies are often ignored. A 1977 study presented to the Eighth World Forestry Congress, Jakarta, 1978, examined the operations of nine logging companies in Kalimantan, Indonesia, and found that "none was leaving the required 25 select crop trees per hectare (10 trees per acre)".

- Indonesia has recently started to require concessionaires to put up a performance bond returnable only after companies have undertaken reforestation programmes.

- Nevertheless, many experienced foresters predict the disappearance of primary TMF in Sumatra and Sulawesi within 5-10 years, and in Kalimantan (Borneo) by 1995 at the latest.

- Half of Peninsular Malaysia's rainforests have been logged in the past 20 years. The Forestry Department forecasts that the remaining forest resources will be exhausted by 1990 at the latest.

- A Fiji forestry official was reported as stating, "the balance of advantage would appear to be strongly in favour of felling and selling as much as possible. If it were possible to do so efficiently it is arguable that it would be desirable to fell all of the natural timber over, say, the next five years."

- West Africa accounts for 15% of TMF timber extraction.

- Up to 5,000 square kilometres (1,950 sq mi) of the Ivory Coast's TMF is logged by timber extractors each year. Replanting, at 30 square kilometres (12 sq mi), is nowhere near keeping pace. So not only will the nation's TMF be finished by 1985, so will the Ivory Coast's role as a timber-producing country. Timber provides a quarter of its foreign exchange.

- Latin America, with the largest and as yet <u>least exploited timber reserves</u>, contributes 18% of world output of tropical hardwoods. The Brazilian Forestry Department wants 400,000 square kilometres (154,000 sq mi) of forest set aside for timber exploitation, and the Superintendency of the Amazon has asked for 800,000 square kilometres (309,000 sq mi) - 30% of all TMF in the Brazilian Amazon - for the same purpose.

- In most cases official plans are academic, as governments lack the means of <u>enforcing their desires</u>.

- Few Latin American TMFs have ever been the subject of longterm management schemes. Contractors have shown no interest in such schemes, and there is little theoretical understanding or practical experience to provide a basis for rational timber exploitation programmes in the region. Quite simply, <u>no one knows how</u> to manage a TMF.

- The ecological effects of logging <u>vary with the methods</u> used. Given the complex ecology of TMFs, the full range of effects associated with different techniques is not fully understood.

- The two main ways of harvesting timber are <u>selective logging and clear felling</u>. Under the former, only trees of an acceptable size and species are harvested. Proponents of the system say it leaves the forest relatively intact, and allows species to regenerate. Opponents say that selective logging severely damages the forest.

- A 1977 symposium on the effects of logging in South East Asia was told that in one Malaysian dipterocarp forest, although only 10% of the trees were harvested, <u>55% were destroyed</u> or severely damaged by falling trees, machinery or the pull of tangled vines; only 35% remained undamaged. Damage is exacerbated by the ease with which a torn limb or cut bark can become infected.

- Critics also maintain that selective logging operations do not leave behind enough seed trees or that they tend to leave only the <u>less well-formed specimens</u>, which will produce poor trees.

- Selective logging also fails to realise the forest's <u>full economic potential</u>, partly because the market will accept only a few well-known species. Southeast Asia relies on 12-15 species for most of its exports; African forest exports are dominated by 10 species.

Roads

Forests are initially <u>opened up by roads</u>.

- This has clearly been the case in Africa. An FAO/UNEP report describes the African experience of the last 20 years: "...opening of logging roads, rush of alien populations using these roads to penetrate the forest, each family settling a few hundred metres from their neighbour in order to secure the largest extension possible from the first clearing. The end result is a <u>gradual denuding</u> of forest areas, in which the many clearings become larger and larger and merge after a few years."

- Brazil's Transamazonian Highway illustrates some of the problems. The Belem-Brasilia Highway, completed in 1959, attracted tens of thousands of settlers. By 1978, the population along the road was 2.5 million, up from 100,000 10 years before. From 1970-78 the territory of Rondonia experienced an annual <u>population increase of 21.3%</u> as a result of the 1969 opening of the Cuiaba-Porto Velho highway.

- In 1970, the government announced plans for a series of highways across Amazonia, in conjunction with an ambitious scheme to settle 500,000 people along new roads by 1975. Within a few years, the programme was tacitly <u>acknowledged as a failure</u>, and the 7,000 families who had been officially relocated (at a cost to the government of $65,000 apiece) along with the thousands more who had followed the highway, were left to live on the inhospitable forest soils without the support which had been promised them.

In western Brazil a "road" - nothing more than a long, thin strip of cleared TMF - has caused severe erosion. (see page 31)

Photo: *Marcos Santilli/Earthscan*

- Another 12,000 unofficial colonisers have been granted <u>legal title</u> to the land they occupy in Rondonia, leaving over 400,000 who have settled along the roads in the area in the position of squatters.

- <u>Bad planning</u> was the chief culprit for this social and ecological disaster.

* The construction of 22,500 kilometres (14,000 miles) of roads was undertaken without any preliminary soil studies. Some routes turned out to be <u>under water for several months</u> of the year. The colonists were unfamiliar with the requirements of the lands on which they settled, and so were their technical advisers.

* Transport and marketing constraints meant that even those settlers lucky enough to find themselves on fertile soil were often <u>unable to sell the surplus</u> they produced. In Cacual, an area of rich volcanic soil, 65,000 settlers were able to raise food crops of coffee, rice, cocoa and maize; but 40% of the 1979 harvest rotted due to lack of markets and transport during the rainy season.

- Damage is caused not only by the settlers which roads bring in, but by the space the roads themselves require. <u>Over 25%</u> of all Brazilian Amazon deforestation between 1966 and 1975 was caused by highway construction, according to official figures.

- Another effect of the highway system was to wreak havoc on <u>96 of Brazil's forest-dwelling groups</u>, whose traditional lands were intersected by the new roads. Half of these groups had experienced little or no contact with the outside world before the roads came, bringing diseases and social problems against which they had no defences.

- A 1981 World Bank study of Brazil noted that deforestation and rapid settlement in the northwest region may cause changes in the environment which may in turn increase the incidence of certain <u>tropical diseases</u>, especially malaria, against which the majority of migrants have no resistance.

- River blindness (onchocerciasis) is another disease that has been spread by opening the forest. Despite warnings, the Northern Perimeter Highway was laid through the only area of Brazil afflicted with the disease. Since then <u>every member</u> of one group of Yanomamo Indians has been stricken with the disease, and nine other previously unaffected tribes are also seriously afflicted.

- A similar tragedy occurred in Malaysia in 1976 when forest clearance resulted in outbreaks of the mosquito-borne disease <u>dengue fever</u>.

- Opponents of the construction of the Darien Gap Highway in Panama, which will complete the Pan-American Highway linking <u>Alaska to Patagonia</u>, point out that it will also link the healthy cattle of North and Central America with the foot-and-mouth epidemics of the south.

Cattle ranching

Many large-scale farmers are concerned not with sustainability, but with maximisation of profit. Short-term schemes suit them, especially if land is priced sufficiently low to enable profits to be made before the land collapses.

- Cattle ranching is the dominant factor in deforestation in Central America.

- More than a quarter of all Central American forests have been destroyed since 1960 to produce beef, 85-95% of which went to the US. This represents less than 2% of total US beef consumption, but has a devastating effect on Central American forests.

- Official figures attribute 38% of all deforestation in the Brazilian Amazon between 1966 and 1975 to cattle ranching (90% of this under a state programme of fiscal incentives); 31% to agriculture (over half state-sponsored); and 27% to highway construction.

- Between 1965 and 1978, government-sponsored ranches cleared 80,000 square kilometres (31,000 sq mi) of primary forest. Another 32,000 square kilometres (12,500 sq mi) has been earmarked for new ranches (mostly in semi-deciduous forests, also on unsuitable soil). These figures may soon be dwarfed by a scheme to allocate 400,000 square kilometres (154,000 sq mi) to agriculture and up to 1.2 million square kilometres (463,000 sq mi) for timber exploitation.

- In Central America, man-established pasture more than doubled between 1950 and 1975, mostly at the expense of primary forest.

- Most ecologists agree that conversion of TMF to pasture is the worst possible use of the land. Constant burning is required to keep down weeds and combat the compacting effect of cattle hooves just to maintain productivity for 5-7 years. After that time the land is exhausted, toxic weeds become the dominant vegetation (resulting in the loss of 10-15% of cattle in the Brazilian Amazon), and the rancher must move on.

- Despite the growth in cattle herds, domestic beef consumption in many Central American countries has declined in recent years. Guatemala's annual beef exports rose from zero to 13,600 tonnes in a decade, while domestic consumption fell 50%.

- Most of the Central American beef exports to the US are used in tinned and pet foods and cheap hamburgers. Its cheapness in the US - less than half the price of domestically produced grass-fed beef - does not reflect the high environmental costs of its production.

- Some 85% of the cattle ranches around Paragominas, a major livestock centre in Amazonia, had failed by 1977, due to soil exhaustion and herd losses. In most cases capital losses were small because of the low cost of buying and running the ranches, and substantial government subsidies and tax exemptions.

Mining

Mining is a <u>minor cause of deforestation</u>, though associated activities, such as road building and the discharge of silt and toxic affluents into rivers, can disrupt a forest more than the mining itself.

- Many TMF regions are rich in mineral reserves.

    * The <u>Amazon</u> has major deposits of bauxite, cassiterite (tin ore), kaolin, manganese, gold and diamonds, as well as the world's largest known iron ore reserves, the Serra dos Carajos, near the Tocantins River

    * <u>Peru and Ecuador</u> have oil

    * <u>Brazil, Bolivia and Gabon</u> have uranium

    * <u>Indonesia</u> has large deposits of nickel, copper, bauxite, tin and coal.

- Many rich mineral deposits coincide with areas which are also the only habitats of certain plant species, so the relatively small amount of deforestation necessary for mineral extraction can have a <u>disproportionately destructive</u> effect on the overall ecological system.

- One of the world's <u>largest known copper reserves</u> is the Cerro Colorado deposit in Panama's western Chiriqui province. Rio Tinto-Zinc (RTZ), which will have a 49% interest in the mine, the cost of which will be about $1.8 billion, is now preparing a feasibility study. The mine, with 1.3 billion tonnes of ore awaiting recovery, is likely to give rise to human and environmental problems.

- The loudest objections have come from the Guaymi Congress, representing <u>7,000 Guaymi Indians</u> who will be directly affected. The Guaymi are subsistence farmers and labourers on plantations and cattle ranches in the area. Most are desperately poor. As a result of contact with the dominant society, one in five Indians has chronic tuberculosis, and three in five children are severely malnourished.

- The fertile areas of Chiriqui have been commandeered by growers of luxury export crops such as <u>coffee and bananas</u>. The Guaymi have not been compensated for or consulted about the assigning of their traditional land to the Panamanian mining company, Codemin.

- RTZ says it is undertaking studies and programmes to limit the impact of the mine on the local people. But environmentalists share the Indians' concern that the 27 million tonnes of earth and rock which will be moved each year will damage the quality of rivers on which the Guaymi depend. An associated hydroelectric project in the Rio Teribe may flood Guaymi lands and damage the river's fish stocks. The Guaymi Congress is especially worried by the social impact of the mine, with 3,000 miners expected to create a boom economy in which <u>drink and prostitution</u> could overwhelm what is left of traditional culture and values.

- If environmental safeguards are not implemented, mines may discharge <u>toxic waste</u> into rivers and streams, and deforestation and excavation may lead to erosion, siltation and landslides.

- In Sabah, Malaysia, 620 families had to be resettled in 1980 when landslides at the site of an open-cast copper mine left more than 1,500 hectares (3,700 acres) of rice paddies under as much as 30 centimetres (12 inches) of mud. Another 50,000 villagers along the River Sugut are endangered by water pollution from the same mine. At their request, Sabah's Environmental Agency conducted pollution monitoring tests which revealed a copper concentration of 2 parts per million (ppm) in the river, <u>10 times the standard</u> amount. High levels of zinc and chrome have also been measured.

CHAPTER FIVE

ACTION AND PROPOSALS

A high priority must be to <u>increase knowledge</u> of TMFs and how they work.

- Satellite monitoring, radar and aerial photography make surveys <u>more efficient and cheaper</u> than ground analysis. Brazil, Thailand, Philippines, Colombia, Venezuela, Peru and Indonesia have recently completed or are about to complete forest surveys. Together they account for half the TMF area.

- FAO's Tropical Forest Resource Assessment project (as part of UNEP's Global Environmental Monitoring System), is using data from US satellites to construct a <u>country-by-country picture</u> of forest cover and of the rate and causes of change.

- Forest surveys, which often cover little-charted territory, can have <u>useful side-effects</u>:

  * a recently completed Brazilian Amazonia survey (Projecto Radambrasil) revealed the existence of a <u>major tributary</u> of the Amazon, and showed that several mountain ranges are mislocated on existing maps

  * Projecto Radambrasil also established that only 2-3% of soils in the area can support <u>continuous agriculture</u> of the kind planned under the government's colonisation programme in the early 1970s.

- A national forest inventory carried out by the General Electric Company of the US for the Philippine government showed that deforestation was far more extensive than had been suspected.

Government officials realised that all accessible dipterocarp forests were in danger of exhaustion by 1985, and an <u>intensive afforestation campaign</u> is now underway in the Philippines.

Putting a value on forests

Many politicians and planners view intact TMF as <u>wasted potential</u>. They see the forest as holding little economic value until it is logged or cleared for farms, roads or mines.

- Although it has become increasingly clear that the value of cleared forest land is rarely as great as expected, little attention has been given to developing analytical models that give full weight to the <u>economic value of intact forest</u>.

- Producing such models might be the single <u>most effective measure</u> that could be taken in defence of TMFs.

- Methods of figuring the true economic value of TMFs would also help decision-makers figure the cost of deforestation. And it might lead to payments of <u>more equitable compensation</u> by those who benefit from TMF resources.

- For example, planners need to know the relationship between <u>tropical agriculture</u> and TMFs. Food supplies for nearly two billion people in the tropics depend on soil and water, the quality of which in turn depends partly on the proper functioning of the TMF ecosystem. The millions of dollars worth of damage caused every year by flooding and siltation as a result of deforestation in all forest areas should also be taken into account.

- The undervaluing of intact TMF benefits those who have concessions to log, mine or otherwise alter the ecosystem. By not assigning economic values to the ecological functions and genetic wealth of the forest, TMF governments are selling their non-renewable <u>natural resources cheap</u>.

- Failing to quantify the economic benefits accruing from intact TMF also means that governments are <u>unable to plan</u> rationally for the future.

- It may make economic sense for a company such as Volkswagen to graze cattle on a 120,000 hectare (296,000 acre) ranch carved out of a TMF, especially as the land is cheap and fiscal incentives are good; but it is the host nation that <u>suffers the long-term loss</u> when companies behave irresponsibly.

- In 1976, two years after the event, Brazilian officials learned from a satellite photograph that Volkswagen had started the <u>largest fire in the world</u>, by spraying 25,000 hectares (62,000 acres) of forest with defoliant and setting it alight, as a cheap and easy way of clearing it for pasture. Only 9,000 hectares (22,000 acres) had been approved for clearance.

- This also raises the question of putting a pricetag on the genetic riches of the forests. Plant and animal samples, collected for scientific purposes and used to create new medicines, crop varieties or other useful products, have traditionally been allowed to leave the country free of charge.

- But these plants and animals, containing valuable chemical compounds, are as much a national natural resource as the oil and minerals under the forest floor.

- Rather than having TMF countries ban scientific collections or wait until they have the capacity to develop the potential of these resources themselves, a system of royalty payments could be arranged as a way of encouraging innovation and ensuring fair compensation to the country of origin.

- Forest dwellers who discovered the now medically important effects of curare (a muscle relaxant), coca (source of cocaine) and cinchona (quinine) have never been recompensed for their contributions to modern medicine.

Research

Scientific ignorance of the TMF ecology is staggering. Dr S.H. Sohmer of the National Science Foundation in Washington DC has said, "We should never lose sight of the fact that we know little or nothing about the way most humid tropical forests are structured and how they function, not to mention the component species and how they evolve".

- Practically nothing is known, for example, about tree growth rates in tropical forests, or about the germination process or light and water requirements of tropical tree species.

- According to Dr Tom Lovejoy of the World Wildlife Fund-US, "In the Brazilian share of the (Amazon) Basin, as far as we can determine, the largest area for which both reliable and published tree survey data exists is a mere 5.5 hectares (13.5 acres)".

- Research in a biological system as rich and complex as TMF can yield not only information helpful in the management of that system, but also information useful to the advancement of scientific understanding of the world as a whole.

- Rational exploitation of forests can be promoted through:

  * basic research institutions, such as Brazil's Institute for Amazon Research (INPA)

  * applied research and experimental field stations such as the Centre for Tropical Agriculture (CATIE) in Costa Rica and the International Institute of Tropical Agriculture (IITA) in Nigeria.

Funding

Tropical research is seriously underfunded in comparison with other fields, and the funding is unevenly distributed. Tropical staple crops are virtually ignored in favour of studies of export crops such as rubber, oil palm and sugarcane.

- Only 4,000 scientists worldwide are working on tropical ecosystems. Half of these are taxonomists, whose concern is to discover and classify the millions of tropical plant and animal species.

- Worldwide there are fewer than 25 scientists competent to undertake or supervise large-scale studies of tropical ecosystems, according to Dr Peter Raven of the Missouri Botanic Garden, himself a tropical ecologist.

- By the end of 1979, the US National Science Foundation had committed just over $2 million of a total budget of $80 million to studies connected with tropical humid forests.

- Total US spending on tropical research, including money spent by universities and government agencies, amounted to about $20 million in 1979.

- A recent US National Academy of Sciences report recommended a doubling or tripling of that figure.

- Global spending on tropical research is less than $40 million, making the US by far the largest source of funding.

- Among international agencies, the United Nations is the major source of finance for TMF research. Between 1974 and 1978, UNEP spent $3 million on basic practical work, establishing land use guidelines, developing forest cover, monitoring projects and advancing inter-disciplinary communication on TMF problems.

- UNESCO's Man and the Biosphere (MAB) programme is one of the most promising research efforts.

- There are now 80 MAB field projects concerned with tropical and subtropical forest ecosystems. Eventually about two billion hectares (five billion acres) of tropical and subtropical forest or once forested lands will be involved in the MAB programme.

- Many MAB projects are located in areas where development projects are, or soon will be, underway (such as Papua New Guinea's Purari Basin or East Kalimantan in Indonesia) in response to the pressing need for research to keep pace with events.

- The projects are concerned with every aspect of tropical forests, from fundamental scientific research to social impact studies. In East Kalimantan, MAB researchers are studying:

* the effects of timber camps on the economic activities of established villages

* the comparison between <u>native shifting agriculturalists</u> and downriver resettlement areas

* the <u>environmental effects</u> of various kinds of land use in the area.

- In San Carlos de Rio Negro in Venezuela's Amazon region, MAB ecologists are conducting research on the functioning of a <u>tropical lowland forest ecosystem</u>, studying the soils, climate, forest structure, zoology, nutrient cycling and response to natural and human disturbance.

- There are also eight MAB "Biosphere Reserves" in TMFs: in the Ivory Coast, Nigeria, Central African Republic, Congo, Zaire, Sri Lanka, Peru and Bolivia. A country declaring such a biosphere reserve <u>loses no national sovereignty</u>, but the area is set aside for scientific research as part of a global cooperative programme.

- <u>The World Wildlife Fund</u>, the largest non-governmental international conservation organisation, supports a number of projects in TMF areas including the successful Operation Tiger in India and the Mountain Gorilla Project in Rwanda and Zaire. It has agreements with some countries to formulate management plans for certain threatened areas, as it has done in Indonesia in conjunction with Survival International.

- The International Union for Conservation of Nature and Natural Resources (IUCN) has published guidelines for development of TMFs which take account of the plight of forest dwellers. Its <u>threatened plants committee</u> is conducting inventories of <u>threatened and endemic species</u> in Latin America, Africa and Asia.

- There are also a number of <u>regional cooperative measures</u> designed to improve knowledge and utilisation of TMFs.

- In 1978, eight Latin American countries signed the <u>Amazon Pact</u>, pledging to coordinate their national scientific and development research efforts. The pact calls for a balance between economic growth and environmental preservation in the Amazon.

- In 1980, <u>Cameroon, Gabon and Zaire</u> ratified a convention on improved management of tropical forests. Six other African countries have also agreed to the convention, and UNEP is helping to finance a regional Centre for Scientific Information and Documentation on Tropical Biology, to be based in Cameroon.

Plants and animals

The "Global 2000" report, drafted at the request of US President Jimmy Carter, estimated in 1980 that one million TMF species could be extinct <u>by the end of the century</u> if deforestation continues at the present rate.

- Such a rapid depletion of genetic material has grave consequences for non-TMF countries: <u>98% of US crop production</u> is based on species from outside the United States.

- Both <u>Brazil and the Philippines</u> have recently experienced the dangers of over-reliance on a few modern crop strains at the expense of the habitats of wild and primitive varieties.

- Brazilian coffee plantations, dominated by <u>a single variety</u>, were badly damaged by bad weather and an attack of leaf rust. Luckily, African forests supplied the genetic material necessary to breed a new resistant strain.

- The Philippines were less fortunate. When several new super-breeds of rice failed to flourish in the islands, agronomists decided to go back to a native strain that possessed the qualities they needed. But the new plantings had virtually <u>eliminated the original breed</u>.

- Thus the point is not just saving species, but conserving the genetic <u>variation and diversity</u> of species. The survival of a rare species may be of esoteric interest only, but the survival of an outlying variation of a well-known species of an economic crop - rice, various cereals, coffee, bananas - may be vitally important.

- With 85% of the world's food supply derived from <u>only eight species</u>, TMFs could make an important contribution to agriculture in terms of new food crops, especially for areas not presently well-served by the temperate zone bias of developed agriculture.

- Of the <u>251 species of tree</u> that bear edible fruit in Papua New Guinea, only 43 are in cultivation.

- UNEP executive director Dr Mostafa Tolba has pointed out that "in 1970 virtually nobody outside of South East Asia had heard of the <u>winged bean</u>. In 1980 this protein-rich legume is being grown in 50 countries throughout the tropics." It was not a botanist or a plant breeder who discovered the value of the winged bean, but the Papua New Guinea forest tribes who have cultivated it for years.

- A new species of citrus was found recently in a Malaysian rainforest. The discovery has great potential in breeding programmes due to the fruit's <u>tolerance of wet conditions</u>.

- The US National Academy of Sciences has pointed to the economic potential of a number of relatively little-known plants, mostly <u>natives of TMFs</u>. They include:

  * cocoyams (several species of Xanthosoma) and taro (Colocasia esculenta), two nutritious <u>potato-like tubers</u>

  * <u>wax gourds</u> (Benincasa hispida) which produce three or four crops a year of a mild, juicy, melon-like vegetable capable of being stored up to a year even in the humid tropics

  * <u>six fruits</u>, including the mangosteen (Garcinia mangostana) of South East Asia and the naranjilla (Solanum quitoense), "the golden fruit of the Andes", found mainly in Colombia and Ecuador.

- In addition, forest dwellers cultivate or gather many thousands of food plants which are still <u>unknown to outsiders</u>. Indonesia alone has an estimated 4,000 such plants.

- The leaves of the Stevia plant have been used for generations by Paraguayan Indians as a sweetener. Now Japanese researchers have analysed its constituents and found a chemical which they say is calorie-free, harmless to humans and tastes <u>300 times sweeter</u> than sugar.

- The US National Academy of Sciences has pinpointed three Amazonian trees as commercially promising <u>sources of oil</u>.

- One of the trees, Jessenia polycarpa, produces an oil <u>similar to olive oil</u>. It is sold in markets in Bogota, Colombia, but is unknown elsewhere. The milky residue from oil extraction makes a good beverage and the seeds are edible.

- Another tree, the Babassu Palm (several Orbignya species) produces a fruit that tastes, looks and smells like coconut, but contains 72% more oil. The oil can be used to produce soap, detergent, starch and general edibles. After extraction of the oil, the leftover seedcake makes a <u>high-protein animal feed</u>. The husk is burned as fuel or converted to charcoal.

- The tropical copaiba or copaifera tree produces 20 litres (five US gallons) of <u>diesel-like heavy oil</u> per tree every six months, according to Nobel prize-winner Melvin Calvin. This oil can reportedly be poured right into a diesel engine, and since 1978 the Instituto Nacional de Pesquisas has run its Toyota pick-up trucks on this fuel.

- The National Academy also recommends the exploitation of Guar (Cyamopsis tetragonoloba), an Asian plant that it calls the most promising <u>source of vegetable gum</u>, with applications ranging from reducing water resistance in fire hoses to thickening ice cream.

- TMFs contain many <u>fast-growing tree species</u> which could be useful in plantations as firewood, timber, animal fodder and protection for deforested areas.

- One of the <u>most versatile</u> is Leucaena leucocephala, an evergreen leguminous tree which grows four metres (13 feet) in its first six months; it can reach almost 10 metres (33 feet) by the end of the second year and 15 metres (50 feet) in six years.

- The leaves are nutritious livestock fodder, and can also be used as green manure: one hectare (2.5 acres) of Leucaena leaves is sufficient to fertilise two hectares (five acres) of annual crops. Because it is a member of the pea family (the Leguminosae) it has the ability <u>to fix nitrogen</u> in the soil, thus improving the quality of the soil in which it is grown.

- Ten years ago the <u>value of medicinal drugs</u> developed from wild species was put at $3 billion per year. Yet the search for useful active ingredients in wild plants and animals is in its infancy.

- There is a constant struggle in the humid tropics between predators and their victims, mostly plants. Many TMF plants have evolved the ability to produce toxic compounds to keep off insect attacks (caffeine, cocaine, reserpine, strychnine) and many insects the ability to digest them safely.

- Curare can kill; it is also one of the most important muscle relaxants in the modern pharmacopoeia, and essential to many surgical procedures.

- Cunaniol, derived from a plant used by Guyanese tribes as a fish poison, may prove to be a great aid in heart surgery.

- As the prices of petroleum-derived synthetic chemicals rise, the attractions of phyto-chemicals (chemicals derived from plants) become greater.

- Over 40% of all prescriptions in the US have as their main or only active ingredient a drug derived from nature.

- In 1974 the US imported $24.4 million worth of medicinal plants.

- Vincristine and vinblastine, two extracts from the rosy periwinkle (Vinca rosea), a native of the West Indian TMF, have since 1960 increased by 400% the chances of successfully treating cases of Hodgkin's disease.

- These two substances, with annual sales of over $20 million, have also given lymphocytic leukaemia patients a 99% chance of recovery and have proved 50-80% effective in fighting several kinds of cancer.

- Dozens of potential anti-cancer agents from tropical forests are undergoing clinical trials. The National Cancer Institute in Maryland, USA, which is carrying out a large-scale screening programme in a search for plants with anti-cancer properties, has said that the elimination of plants from tropical forests would be a major setback to the fight against cancer.

- A survey of 1,500 Costa Rican plants showed that 15% might contain anti-cancer properties.

- Contraception is another area of research which depends heavily on TMF resources.

- Without the Mexican Yam (Dioscorea composita), which yields virtually the entire world supply of diosgenin (from which testosterone, progesterone and other sex hormones can be produced), there would have been no birth control pill.

- Diosgenin can also be used to produce cortisone, widely used in treating rheumatoid arthritis, allergies, sciatica and other skin diseases.

- WHO has a task force on Indigenous Plants for Fertility Regulation. One possible candidate is the greenheart tree (Nectandra rodiaei), a native of the Amazon, which researchers hope may be the source of an effective female contraceptive.

44.

- There are countless other areas of potentially useful research. For example, a study of plants that have developed a resistance to certain insects could provide information about mechanisms or substances relevant to the development of insect-resistant crops.

- Plants may also produce substances useful as externally applied pesticides.

- Certain members of the pea family in TMFs produce insect repellant substances - "rotenoids" - in their roots; from these scientists hope to develop effective and safe insecticides.

- The other side of the equation - insects that can cope with toxic compounds - also has exciting possibilities for exploitation in medicine and agriculture.

- Research into TMF wildlife management would also be productive. It scarcely exists at present, but is economically and ecologically feasible.

- Virtually all TMF species are hunted for meat by forest dwellers. In some forest areas of Nigeria, wildlife provides over 80% of protein consumed.

- A survey of 150 square kilometres (58 sq mi) of Malaysian lowland rainforest discovered 76 different species of mammals. When the forest was converted to rubber and oil palm plantations, only 13 species were left. All of these, apart from two species of rat, were dependent on nearby forest.

- The kouprey, an ancestor of one of Asia's commonest cattle breeds, the humped zebu, lives in forests on the border between Thailand and Kampuchea. It could expand the genetic base of cattle breeding throughout Asia, but it is a highly endangered species after 15 years of warfare in the region.

- Southeast Asian forests are home to a number of other wild bovids which could be useful as domesticated animals.

- Fish resources, too, have scarcely been tapped.

  * The Tonle Sap River region in Kampuchea is 10 times more productive than the best Atlantic fishing grounds

  * The Amazon and its drainage basin have received virtually no commercial attention, yet they contain between 2,000 and 4,000 fish species, only 60% of which have been scientifically identified.

- Nowhere are tropical rivers and lakes managed intensively in a way comparable with the fishing areas of developed nations.

- This is partly because scientific understanding of tropical rivers is rudimentary. Peter Raven, a leading tropical ecologist, has pointed out that when we attempt to establish freshwater fisheries or otherwise manage tropical freshwater systems, "we are attempting to manipulate a system in which only somewhat over half of the elements have even been registered, much less understood".

- <u>Forest animals</u> are also important for medical research:

  * the US imported <u>34,000 primates</u> from TMFs in 1977

  * <u>chimpanzees</u>, of which there are only 50,000 in the wild, are the only animals on which anti-hepatitis vaccine can be tested

  * an important anti-cancer vaccine is extracted from the <u>cotton-topped marmoset</u>, a species vulnerable to cancer of the lymphatic system.

Reserves

Brazil has a law requiring that <u>half of every forest</u> holding be left unfelled; but there is nothing to stop the owner from clearing half and selling the rest to a new owner, who is then entitled to clear half of his holding, and so on. However, even if the spirit of this provision were always respected, the resultant checkerboard pattern of forest cover would be of questionable value in the conservation of species.

- One estimate of the <u>minimum critical area</u> needed for a TMF reserve is around 2,500 square kilometres (965 sq mi). But there have been few long-term studies of tropical forest reserves, the process of ecosystem decay is still little understood, and estimates of minimum reserve areas vary widely.

- The <u>only published example</u> is of the Garden Jungle in the Singapore Botanical Garden where, by 1977, a 200 by 300 metre (655 by 985 feet) plot of formerly continuous dipterocarp forest retained only one or two dipterocarps, no seed parents and no seedlings.

- A new concept in protected areas is the theory of <u>biogeographical islands</u>. According to this theory, when cold, dry glacial conditions prevailed in the Pleistocene era, tropical moist forests persisted only in fragments, retreating to form "islands", isolated from one another for long periods during which new species evolved.

- As conditions improved the forest <u>once again spread out</u> to its present range.

- These islands, which have been identified only for a few species of butterflies, birds, trees and reptiles, are believed to have remained centres of <u>exceptional biotic diversity</u>: if protected they would save proportionately more species than any other TMF areas.

- By identifying and protecting such "<u>Pleistocene refugia</u>", it may be possible to save 75% of Amazonian species on only 5% of the land. Now only about 1.5% of TMF is protected, in many cases inadequately.

- Most scientists believe that a well-chosen network of reserves capable of supporting most TMF species would require <u>only 10% of TMFs</u>.

- In order for these reserves to receive adequate protection it might be necessary to surround them with buffer-zone "<u>green belts</u>". These would be areas of forest preserved for watershed protection, collection of certain valuable plants or animal species, education, scientific research, tourism and recreation.

Alternate sites

One way of reducing pressure on TMFs would be to site the agricultural or settlement projects threatening the forests on more suitable land. Alternative sites have been overlooked because of the <u>deceptively fertile appearance</u> of TMFs and because the real value of intact TMF has not been taken into account.

- Although much TMF soil is nutrient-poor and unsuited to permanent agriculture, there are patches of high fertility areas of natural grassland or savannah, and floodplains capable of <u>supporting settled agriculture</u>. There are also millions of hectares of secondary or degraded primary forest which should be the focus of any new agricultural or industrial projects.

- Secondary forests abound in TMF regions, but entrepreneurs often avoid locating new developments there because they can make an <u>initial windfall</u> by selling the valuable hardwoods felled in clearing a primary forest tract.

- Dr Tom Lovejoy, a tropical biologist and vice-president of the World Wildlife Fund-US, reckons that the amount of <u>already deforested land</u> in TMF countries would be sufficient to meet all their timber and pulp needs if a way of sustaining productivity could be found.

- The Amazon, for example, has 50,000-100,000 square kilometres (19,000-39,000 sq mi) of <u>relatively fertile soils</u>, and 150,000 square kilometres (58,000 sq mi) of savannah (cerrado).

- Brazil also has 1.6 million square kilometres (6.2 million sq mi) of cerrado to the south of Amazonia, an area three times the size of France, with soils that are <u>tougher and better suited</u> to cattle raising than those of the forest. Also, it is much easier to farm.

- Many foresters and ecologists urge the deflection of development from Amazonia to the seemingly bleaker but <u>potentially more productive</u> cerrado.

- According to the World Bank's Office of Environmental Affairs, the Brazilian cerrado is closer to established markets; offers less risk of erosion, pests or disease; and has adequate rainfall, social services, infrastructure and power supply. Soil fertility is low, but response to fertiliser is better than on

Amazonian soils. The cerrado's ecosystem is better understood than TMF, and it supports fewer species and is simpler, so there is <u>less risk of loss</u> in altering the ecosystem.

- Floodplains, made fertile by deposits of alluvial soil, are important as feeding grounds for many freshwater fish species, so it would be a mistake to drain the "varzea" or floodplains and replace existing vegetation with annual crops. But <u>careful planning</u> should enable parts of the floodplain (which includes extensive areas of non-forested swamp) to be exploited for fish, turtles, timber, rubber, jute, beans and rice, while leaving strips of forest along watercourses.

- Rice is the most widely planted crop in Amazonia, covering 200,000 hectares (494,000 acres) of upland forest. Yields average under one tonne per hectare (0.4 tons per acre), and the soil is exhausted after a few years. Floodplain rice can yield 10 times as much <u>on a permanent basis</u>.

- Given the peculiarities of TMF environments, all proposals for TMF development should be subjected to soil <u>productivity and sustainability</u> analyses, to determine the area's potential to support the proposed enterprise.

- The most commonly used land classification system was developed by the US Soil Conservation Service for use <u>in temperate regions</u>. There is a desperate need for a simple, inexpensive method of land-suitability classification appropriate to TMF conditions.

- Such a system was devised recently for use in Venezuela's 160,000 hectare (400,000 acre) Guanare River watershed. The analysis was completed in two man-years, at a cost of <u>28 US cents per hectare</u> (11 cents per acre). Information on climate, soils, geology, vegetation and water resources was used to determine the ecological, economic and cultural suitability of various areas for grazing.

Agroforestry

Agronomists and foresters are beginning to agree that the agricultural approach most likely to succeed in TMFs is one which mimics the <u>diversity and complex interdependence</u> of the TMF ecosystem.

- Peasant farmers have long achieved this by planting a variety of crops, often <u>in conjunction with trees</u>, shrubs and domestic animals.

- In Borneo, Chinese farmers interspace cereals, pepper, rubber and at least 10 varieties of vegetables with fishponds and grazing animals, a system which enables them to farm permanently <u>on deficient soils</u>.

- The Lua hill farmers in the forests of northern Thailand cultivate <u>75 food plants</u>, 21 medicinal plants and 30 others with ceremonial or practical applications.

48.

- In Yurimaguas in the Brazilian Amazon, experiments carried out on land cleared by farmers from secondary forest with poor soils showed that the use of kudzu (Pueraria phaseoloides) as a green manure (mulch) supplies enough nutrients to support production of soybeans, peanuts, cowpeas and upland rice. Yields were 80-90% of similar plots given heavy doses of inorganic fertiliser. Even without the green manure, a five-crop intercropping produced 30% greater output than if the same five crops were planted as monocultures.

- Agroforestry and silvi-agriculture are phrases which cover a wide variety of ways and theories of intercropping plants and trees. In some cases, such as the taungya system, which originated over a century ago in Burma and is now widely practised in the tropics, especially in Nigeria, food crops are secondary to the main purpose of the exercise, which is to establish a healthy forest.

- The Nigerian taungya system does not solve the problem of food production for the poorest, since after the forest canopy closes and crops can no longer be grown there, the farmer must move on. But it is an effective and economical way of establishing forests.

- Nigerian plantations begun through taungya cost US $200-300 per hectare ($80-120 per acre) compared with $800 per hectare ($320 per acre) with direct planting.

- A forest village scheme established in 1968 by the Thai government's Forest Industries Organisation uses the taungya system to reforest degraded land.

- The establishment of forest villages is preceded by a public information campaign and consultations with local leaders. Although electricity, roads, schools, water, social services and technical assistance are provided, it is difficult to persuade shifting cultivators to settle and stay in forest villages. Improving land tenure and incentive payments has increased the rate of permanent settlement in the new villages.

- Every village is allotted an area of forest large enough to be replanted and harvested over a 60-year cycle (most plantations are teak, a long-rotation tree that occurs naturally in monospecific - that is, teak-only - stands). Cash crops may be planted along with tree seedlings. Each family also has a small plot of land on which to build a home, and maintain a garden and raise poultry or pigs. The villagers are paid for their plantation work and allowed to sell the cash crops. The average monthly family income in 1978 was US $30-50. In the project's first decade, 19,000 hectares (47,000 acres) were reforested.

- The World Bank is funding a project in the Philippines in which forest farmers grow timber as a cash crop. The programme encourages farmers to devote 80% of an average 10-15 hectare (25-37 acre) plot to trees to provide raw materials for industry; the other 20% is for food crops. But while such programmes may produce trees, they destroy the genetic diversity of TMFs.

- The principal species grown are Albizzia falcataria (known locally as Molucca Sau) for pulpwood and Leucaena leucocephala for fuel and leafmeal on 4-8 year rotations. By March 1978, some 3,400 farmers were growing 16,600 hectares (41,000 acres) of Albizzia. The programme requires the farmers to sell all their timber to a local company, and the success of the project is closely related to the <u>expertise and enthusiasm</u> with which the local companies and the technical experts assist the farmers.

- According to Dr Harald Sioli, director of the Tropical Ecology Department of the Max Planck Institute of Limnology in West Germany, the most promising project designed specifically to <u>stabalise agriculture</u> rather than create forests is being conducted by German engineer Ernesto Rettlebusch.

- On his 80 hectare (198 acre) farm in the deforested Zona Bragantina in the Amazon, Rettlebusch has 20,000 pepper plants, 20,000 chickens, 500 pigs and 200 cattle. The animal manure is applied to the plantation where it is more effective and longer-lasting than mineral fertilisers. Most farm expenses are covered by the sale of animal products - meat, milk, eggs; pepper is a cash crop; other crops provide the family's food.

- The project is still experimental, with more elements and different mixes to be tried out, including fish-farming and various intercroppings. But other farmers are already beginning to imitate Rettlebusch's combination of intensive agriculture and animal husbandry, with good prospects for the eventual rehabilitation of the area.

- Another promising development, this time in dairy farming, has occurred in the highlands of Costa Rica. A local variety of the alder tree (Alnus acuminata) has been planted in intensively managed pastures, which in turn are divided into small enclosures through which the cattle are rotated every 20-30 days.

- Fertilisers are supplemented by the nitrogen added to the soil by the alders. The wide spacing allows the grass to thrive. The trees are harvested on a 15-20 year rotation, and are highly valued for construction and firewood. It has been estimated that about 60,000 hectares (148,000 acres) in this region could be devoted to such a system, and a programme to encourage farmers to plant more alder is underway.

Plantations

Because the yield per hectare (acre) in tree plantations is so <u>much higher</u> than in selective logging - up to four times the amount of usable wood produced by natural forests - many foresters would prefer to harvest plantation timber.

- Plantations are not necessarily monocultures, but they tend to be restricted to <u>relatively few species</u>.

Land clearance for a housing area on industrialist Daniel Ludwigs's River Jari estate, an area half the size of Belgium cut out of primeval rain forest. (see page 51)

(see page 51)

Photo: Peyton Johnson/FAO

- Drawbacks include uncertainty about the ability of TMF soil to support limited-variety plantations over a long period; their susceptibility to pests and diseases; unsuitability as a habitat for TMF flora and fauna; the loss to local forest dwellers of economically important forest species; and the questionable capacity of plantations to protect the watershed adequately.

- But successful plantations on suitable soil, perhaps with species that can reclaim degraded forest land, could help reduce pressure on remaining TMF and provide a continuing supply of industrial and fuel wood.

- Some of the better tropical soils have supported plantations of rubber, coffee, cocoa or other tree crops for 50 years. Since

1973 Indonesia has sown elephant grass (Pennisetum purpureum) under teak and mahogany plantations. The grass is cut and sold to ranches, but no cattle are allowed in the forest.

- The biggest plantation of all is the $800 million project half
the size of Belgium on the Jari River in Brazil owned by US
industrialist Daniel K Ludwig. Bought in 1967, it ran into a
series of problems and reportedly has been put up for sale. In an
operation on the scale of Jari the most serious effects can be
social and economic rather than ecological. Ludwig planted over
100,000 hectares (247,000 acres) of fast-growing Gmelina arborea
and slash pine (Pinus caribaea) trees, started herds of water
buffalo and cattle, and a rice project.

- But recently harvesting had not started on the pine, and
gmelina had been harvested only since early 1979, so there was no
data to indicate whether the plantation would be sustainable
after the first or second rotation. Also the soil turned out to
be less fertile than expected and the Asian gmelina trees did not
do as well as expected, according to a report by Reuters news
agency.

Logging

Selective logging could be made ecologically and economically
sounder if the value of the damaged trees left behind were added
to the costs of the operation. This would encourage the use of
techniques such as the removal of felled trees from dense forest
by helicopter, a practice used in some North American forests;
severing vines from trees about to be felled; leaving enough seed
trees and maintaining a post-logging regeneration programme.

- A 1977 International Labour Organisation study in the
Philippines showed that for many operations manual felling cost
less than mechanical methods. Manual methods are also more
labour intensive and easier on the soil.

- Clear-felling has been made more attractive by the advent of
"any tree, all tree" wood chip processing, using almost all
species. Advocates claim that clear-felling small areas of the
forest, perhaps in narrow strips, leaving mature forest on
either side, allows natural regeneration to take place and is
less disruptive of the entire ecosystem than an extensive
selective logging operation.

- Not enough is known about natural regeneration to determine the
best way to encourage it after clear-felling. Using bulldozers
for clearing, for example, may compact the soil and inhibit its
ability to regenerate. On the other hand, it may be possible to
grow food crops for the first years after an area of forest is
cleared. Any advantages that may attach to clear-felling depend
on a long-term programme of regeneration management or, in
appropriate circumstances, plantation establishment.

- Logging waste could be greatly reduced. The Washington-based
Conservation Foundation estimates that a typical logging
operation leaves behind on the ground one-third of the wood.

Japanese contractors. The turbine blades of the hydroelectric plant were cracking due to the water's high acidity, a result of flooding the reservoir without first clearing the valley.

Forest dwellers' rights

The World Bank report "Economic Development and Tribal Peoples" identifies their four fundamental needs in the face of pressure from the outside world:

* recognition of territorial rights

* protection from introduced disease

* time to adapt to the national society to the extent they desire

* self-determination; that is, appropriate conditions to maintain their culture and ethnic identity to the extent they desire.

- A key part of the difficulties encountered by traditional forest peoples arises from their uncertain legal status.

- Tribal rights are affected by forestry policy, laws covering religious freedom, compulsory purchase orders and the attitudes of the agencies charged with responsibility for tribal affairs:

* the Indonesian constitution guarantees freedom of religion for believers in "One Supreme God", which excludes the animistic beliefs of many of its forest groups

* in Colombia the Catholic church shares with various government agencies responsibility for Indian affairs

* Bolivia, where over 75% of the population is Indian, has assigned responsibility for the education, medical care and general welfare of its forest-dwelling Indians to the North American Summer Institute of Linguistics, the largest Protestant mission in South America.

- Missionary groups sometimes have the personnel, equipment and knowledge of languages, which many governments lack, to enable them to make contact with and work among forest people.

- But many observers are disturbed by governments which delegate responsibility for citizens to a private organisation with its own interests. They question whether missionaries, whose purpose is to convert people to another way of seeing the world, can avoid damaging the culture on which a tribe's identity depends.

- Some anthropologists and champions of aboriginal rights maintain that in the absence of qualified, disinterested mediators, forest tribes should be left alone.

- However, highways, ranches, logging companies, prospectors and even scientists will continue to bring the dominant culture to aboriginal lands. And though some cultures are fiercely independent, many are receptive to new ideas and goods.

- Forest dwellers need medicine and education so that they are not overwhelmed physically and culturally by their first encounters with the dominant society. Threatened cultures should be alerted to the drawbacks of the Western way of life, and to the problems experienced by other forest tribes who have allowed their cultures to lapse.

- They must also be made aware of the effects of their accepting various aspects of outside culture, such as firearms, which require new hunting patterns if the wildlife population is to be sustained.

- The colonisers and the dominant society at large must be informed of the values and worth of aboriginal cultures. Governments and individuals might be less eager to force forest peoples to conform to the dominant culture if they realised the value of their skills and knowledge.

- On the most basic level, forest tribes have a holistic view of the world, the wisdom of which is just beginning to be realised by ecologists, economists and a few development specialists.

- "Tribal knowledge encompasses the ecosystem in its entirety, of the interdependence of floral and faunal species, of the specificities of micro-zones and their interfaces, of seasonal and longer term variations in plant and animal life, reproduction, growth, movement and productivity: these aspects of tribal knowledge are almost always ignored or disregarded", noted the World Bank report on tribal people.

- "This is in large part because of lack of perception by non-tribal people, combined with difficulties of communication and disdain with which tribal knowledge is often regarded by nationals", it added.

- The now-fashionable notion of integrated rural development is a step closer to the "primitive" concept of life in which the cultural, spiritual and psychological as well as the physical and economic aspects of life are recognised as influencing one another.

- On a material level, the integrated cropping techniques of forest dwellers and their knowledge of forest products are pointing the way for modern agronomists who want to learn how best to exploit the complex TMF ecosystem.

- Forest farmers have developed efficient systems of erosion control, nitrogen enrichment, ground water exploitation, slope management, fertilisation, plant sheltering, drainage, intercropping, forest regeneration, weeding, irrigation and wildlife management.

- The World Bank paper on tribal peoples noted that the settlers taking part in a Swiss-financed project on the Yavari River in Peru often run short of food between outside deliveries because they cannot grow enough to feed themselves. Project officials and colonists usually <u>purloin the needed food</u> from the local Matses (Mayoruna) Indians.

- The Indians "live in precisely the same environment, but even with primitive tools and <u>no outside inputs</u> they thrive enough to tolerate the demands of the much better equipped colonists", the paper said.

- The World Bank report listed five pieces of evidence to back up the assertion that "tribal people can manage sustainably the tropical wet forest ecosystem, and that <u>non-tribal people cannot</u>":

  * tribal people have done so <u>for millenia</u>, and, where left alone, they continue to be successful

  * TMF environments are, by and large, inhabited by tribal people effecting <u>no harm to the ecosystem</u>

  * farming techniques adopted by outsiders - <u>liquidating the resource</u> for short-term profits - force them to move to another tract, eventually ruin the resource base and promote population growth to exceed carrying capacity

  * successful examples of TMF colonisation by outsiders are either "<u>vanishingly rare</u>" or depend on outside capital for energy-intensive and never-ending inputs (petroleum, pesticides, fertilisers)

  * where non-tribal people are cut off from outside sources of food, <u>they starve</u> unless helped by local tribal people.

- The 1978 act establishing Indonesia's Ministry for the Supervision of Development and the Human Environment acknowledged the advanced state of <u>indigenous environmental management</u>.

- The knowledge of forest dwellers about useful TMF plants and animals has already been discussed.

- A number of <u>international agreements</u> cover the rights of tribal peoples to their traditional lands, and these may provide a useful lever against governments which are slow to protect groups of forest dwellers.

- The International Labour Organisation Convention on Tribal and Indigenous Peoples, which has been ratified by 26 countries, stipulates that signatories have a legal duty to adopt "<u>special measures for protection</u> of the institutions, persons, property and labour of indigenous populations".

- Article II of this Convention states that "the <u>right of ownership</u>, collective or individual, of the members of the populations concerned over the lands which they occupy shall be recognised".

Self-help

Forest peoples can also usefully draw on each other's experiences, and form alliances with other interested groups.

- Xavante leaders have joined with other Indian tribes and pro-Indian groups to bring pressure on the Brazilian government to improve its treatment of Indians and, most importantly, to assign them their own lands.

- They travel to the capital to lobby FUNAI (the organisation responsible for Indian welfare) and other agencies, give interviews and publicise their claims skilfully.

- One Xavante leader who was refused entrance to a government office because he was not wearing a conventional suit and tie, announced that Brazilian officials visiting Xavante villages would be required to wear penis sheaths, feathers and body paint.

- The Xavante have changed with the times. They practise stable agriculture with modern tools, speak Portuguese and are able negotiators and organisers for the Indian cause. But they do not wish to merge into Brazilian society.

- Commented an Xavante chief on a lobbying visit to Sao Paulo, "There is a lot of bureaucracy here in the cities. Here you live like a madman, have a lot of problems in the mind without knowing with whom you can solve them. Here in the cities nobody lives in freedom."

- They are determined to continue as Xavante and believe that secure land tenure is essential if they are to determine the progress of their own development.

- The importance of using modern techniques to defend their land and culture has been recognised by many tribal groups.

- A group of Bolivian Indians have made a film documenting their own position and encouraging others to resist deculturalisation and expropriation of their land. The film ends with these words: "The Ayoreo have entered a new phase. Now we are no longer frightened of television, radio and the papers. We use them and know that we can protest. We are Bolivians and we are seeking allies."

- The Gavioes of Brazil have maintained their cultural integrity while entering into a relationship with the dominant culture on equal terms, despite their appalling treatment at the hands of white men since contact was established.

- After half a century of resisting contact with non-Indians, the Gavioes were finally "pacified" in the 1950s. Permanent contact entailed the loss of 70% of their people to violence and disease.

- In 1961 one of the two remaining Gavioes villages contained only eight people, and a visiting anthropologist, Roberto Da Matta, assumed that their end as an independent culture was at hand. Ten years later, a deputation from Survival International

found "every single woman and child was suffering from what
appeared to be whooping cough, and all were sitting dejectedly
with running noses, coughing ceaselessly. When asked what they
were suffering from, a FUNAI representative said, 'Nothing. It
is the normal state for Indians to be in.'"

- But by 1976, the Gavioes population had almost quadrupled to
108. This was despite severe dislocations throughout the period.
One of their villages was moved twice by FUNAI to make way for
the Tucurui Dam, and the other was split in half in 1968 by the
Maraba-Belem road. FUNAI has repeatedly promised the Gavioes a
52,000 hectare (128,500 acre) reserve, but it has not yet been
demarcated.

- Most Gavioes now speak Portuguese and are willing to accept
change. But they are determined to retain their cultural
independence and are prepared to fight for their rights.

- Since the mid-1960s they have developed a profitable trade in
Brazil nuts, earning over $50,000 a year and employing 20
Brazilians. In August 1977, the Gavioes dismissed the FUNAI
representative from their village and took over the financial and
marketing side of the business. Previously they had received
less than 20% of the market value of the nuts they collected.
They have also negotiated a Bank of Brazil loan to expand the
business.

- The Gavioes still face problems from outsiders. In June 1980 it
was announced that Electronorte, the company developing the
Tucurui Dam, had agreed to pay the Gavioes $830,000 for the right
to cut down 300 hectares (740 acres) of forest land for the
construction of electricity transmission lines.

- This works out at US $830 for each Brazil nut tree, which takes
decades to mature, and which yields a very valuable crop. It is
less than half of the amount demanded by the Gavioes, and
certainly much less than forfeited future profits.

- The Txukarramae live in Brazil's Xingu National Park, having
been resettled in 1971 to make way for the routing of the BR80
highway through their traditional territory.

- The road also paved the way for outsiders, mostly poor whites,
to enter Txukarramae lands. Since then, the Txukarramae have
been fighting to regain control of their land.

- In 1979 their chief, Raoni, issued a warning: "From now on
anyone who trespasses on Indian territory will die...We will
fight and maybe we will die, but we are dying slowly anyway."

- In August 1980, Raoni and his warriors kept their pledge. In
two incidents they attacked white settlements on Txukarramae
land, killing 31 adults and children.

- Luis Carlos Silva, a rancher for whom some of the settlers
worked, said: "The United States solved this problem (in its own
West) with its army. They killed a lot of Indians. Today
everything's quiet there and the country is respected around the
world."

- In Indonesia, the World Wildlife Fund (WWF), the International Union for Conservation of Nature (IUCN) and Survival International have initiated projects to protect the forests on Siberut, one of the Mentawai Islands, and to help the people there adjust to change in their environment and their lives.

- Until recently, the Mentawaians followed a complex and stable way of life based on hunting, fishing and shifting cultivation. Extended families based on common ancestors live in long-houses.

- The Mentawai system of shifting agriculture does not involve burning. Crop debris is left to release nutrients gradually by its slow decay and to provide shade for the early crops. Thus in time a more natural forest regenerates because the tree stumps and the mycorrhizal fungi have not been destroyed by fire.

- These "domestic forests" are good homes for wildlife and sources of medicines, fruits, etc. The main crops are sago palms, bananas, and taro. Sago palms give a better yield per labour input than virtually any other crop.

- The Mentawai people collect forest products, but only one tree, the Shorea or Sal, is cut. This is to make canoes, their main form of transport. Hunting with bows and arrows is a social activity surrounded by religious taboos that function to protect wildlife stocks.

- Fifty government and mission villages have been established in the islands in the past 60 years, most of them since 1960. Long-houses are discouraged by the authorities, who prefer large compounds of nuclear families centred on the church. (The Dutch colonial authorities banned "heathen" religions and gave Mentawaians three months to choose either Islam or Christianity.)

- One result of these outside influences on their traditional way of life has been a dramatic population increase in the past 20 years, averaging 3% a year.

- Rice, a more "civilised" crop, is replacing sago palms, though more labour is required for less yield. There is less time for hunting, and because of the concentration of villages, the area hunted is smaller and over-exploited. Traditional taboos that prevented over-hunting are being forgotten. As a result of the poor rice harvests, villagers depend increasingly on store-bought goods.

- There is a new system of land tenure, too. The people own the land but the Forestry Department has rights over the rest and pays no compensation to the villagers for the loss of their only good canoe tree or for any other forest products.

- Over 90% of the land area of the Mentawai Islands has been leased for logging.

- Logging not only depletes what was once an important capital resource, it silts the rivers, making navigation more difficult and reducing the fish catch. Employment by logging companies is marginal because the islanders, concerned with trading their crops and having no experience of working to other people's timetables, are not considered desirable employees.

- WWF and IUCN have designed a management plan for the Mentawai Islands at the request of the Indonesian government.

- The plan calls for a three-tier system:

  * a strictly-controlled, 500 square kilometre (190 sq mi) nature reserve, for which the government has already cancelled all logging concessions

  * a 1,000 square kilometre (385 sq mi) traditional use zone which will act as a buffer zone for the nature reserve, and in which commercial timber extraction will not be allowed

  * a 2,500 square kilometre (965 sq mi) development zone, containing all the villages and areas of greatest agricultural potential.

- Survival International's contribution includes a scheme to offset the reduction in animal resources by introducing new fishing techniques and improved varieties of sago palm.

- The government has adopted the WWF/Survival International project as official policy, but the logging concessions for the traditional use zone have not yet been cancelled.

Forestry institutions

"Though every underdeveloped country now has a forest service, these services are nearly all woefully understaffed and miserably underpaid. Because they exist, exploitation is facilitated. Because they are weak, exploitation is not controlled", according to Jack Westoby, a former FAO forestry official.

- Most TMF conservation plans and programmes, no matter how good they appear on paper, are meaningless, because few governments have the manpower and expertise to administer them. The running of government forestry departments, and the enactment and enforcement of forestry legislation are the most difficult and overlooked aspects of national forest management.

- Most of the world's forestry management practices evolved in temperate regions, but many TMF species are less adaptable to environmental disruption than temperate species.

- State forestry departments have tended to be overshadowed by other bureaucracies, particularly in Latin America, which has the most ineffective forestry institutions of any developing region, according to a 1978 study by the International Institute for Environment and Development (IIED).

- They are not viewed as prestigious institutions and are under-funded, inadequately staffed, powerless and often lacking a coherent forestry policy.

- Around Altamira in Para State, for example, the Brazilian Forestry Institute (IBDF) has responsibility for 200,000 square kilometres (77,000 sq mi) comprising tens of thousands of colonists, numerous ranches, and the city of Altamira itself. Its duties are to enforce the 50% rule (which forbids the clearing of more than half of any landholding), control the export of skins and forest animals, license all timber extractors and sawmills and - since 1979 - to see that no colonist fells a tree without a government-issued licence.   The IBDF staff in the Altamira region consists of a handful of forest officers and guards.

- Paraguay, whose substantial but unsurveyed forests provided timber exports worth $26 million in 1975, has fewer than 15 foresters in its National Forestry Service.

- Paraguay's model land law, categorising all forests for production, protection or "special use" (scientific, educational or historic) exists only on paper, while logging and forest clearance for grazing are virtually uncontrolled.

- Costa Rica has one of the most progressive wildlife protection and parks programmes in Latin America, but not enough personnel to ensure the enforcement of regulations.

- Venezuela, whose wildlife laws are also excellent, is in a similar position.

- A key problem, says Westoby, is that forest services are "largely concerned with assuring and facilitating a steady outflow of woody raw material to the rich countries. Of the new revenues generated, woefully little has been ploughed back into forestry, either into management, into regeneration or new planting, or into research."

- Westoby asserts that "in precious few countries have the energies of foresters been bent upon helping the peasant to develop the kind of forestry that would serve his material welfare.   This is why there are so few village woodlots and fuel plantations.   This is why so little work is done on forage trees, fruit and nut orchards.   This is why more and more watersheds have become denuded, so that the flood and drought oscillations which spell calamity for the peasant become more extreme.   This is why forestry has been invoked so rarely to reclaim or rehabilitate land.   This is why so few of the many possible agroforestry combinations have been actively explored and developed.   This is why so few industries have been established which are specifically geared to meeting real local needs."

- Many forestry regulations are ineffectual, unenforceable or downright harmful.

- Under the Habitat Improvement Scheme, Indonesia allows its largest national park, the over 8,000 square kilometre (3,000 sq mi) Gunung Leuser reserve, to be selectively logged by commercial timber companies. Officials say the system is designed to provide elephants with the secondary forest in which, like gorillas and many other forest animals, they prefer to forage.

Illegal deforestation in North Sumatra's Gunung Leuser Nature Reserve,
Indonesia's largest national park. Indonesia allows "selective logging" in
the park, on the ground that the elephants favour secondary forest. Scientists
have criticised the scheme. (see page 61)

Photo: Alain Compost/WWF

- Scientists point out, however, that if elephant welfare is the
object, it would be better achieved by logging one-hectare
patches every half kilometre (2.5 acres every third of a mile) or
so to provide the truly <u>mosaic pattern</u> preferred by elephants.
Even if depletion of the entire forest did favour elephants, it
would do so only at the expense of the many forest species who
depend on the intact ecosystem.

- In Africa, according to a UN project undertaken as part of
UNEP's Global Environmental Monitoring System, "a <u>huge effort</u>
<u>remains</u> to be made" in natural forest management.

- As in Asia and Latin America, however, little will be possible
without the agreement and participation of local populations,
which will require a large <u>propaganda and extension</u> effort.

- This may require a reorientation in the outlook of foresters in
TMF countries, who frequently regard their forests as

"fortresses" to be protected against local people. Comments Bruce Ross-Sheriff of the US Office of Technology Assessment: "only the mavericks go out and work with the peasants. They (the majority of foresters) think of them like gypsy moths, to be kept out of the forests."

- If wood and water resource wastage is to be avoided, logging of new forest areas must be carried out within the framework of a rational land-use plan, and with harvesting regulations and forecasts of regeneration and reconstruction of logged-over forests. This all requires specialist staff.

- Similarly, the FAO has said that protective management of threatened forests is generally cheaper and ecologically superior to plantations. But this, too, requires forestry institutions with sufficient staff and means.

International agencies

The World Bank's 1978 policy paper on forestry did much to stimulate interest in forestry as an important aspect of rural development.

- The document recommended that the Bank shift attention from large-scale industrial timber projects to fuelwood and small-scale activities. The Bank plans to lend $1 billion to firewood schemes in 1980-85.

- Village forestry, watershed protection, and national institution building are the cornerstones of the new forestry policy. Through its cooperative programme with FAO, the Bank will respond to national requests for forestry training funding, and is also willing to help start new forestry training institutions.

- Until recently, the FAO showed little concern for community forestry, but lately has been preparing a major programme, attempting to quantify the benefits of watershed management.

- Of the development banks, the World Bank has the best system of environmental evaluation of proposed projects, although it has been criticised as being largely a public relations exercise.

- The Bank's Office of Environmental Affairs, with only a small staff, is responsible for screening over 60 projects a month. This over-burdened office is also responsible for monitoring the environmental effects of completed projects.

- The Bank has no effective system for ensuring that ecological considerations are introduced at the outset of all development projects. Many projects are subjected to environmental analysis at too late a stage for the necessary modifications to be easily made, with the consequence that environmental considerations are

often seen as obstructions rather than as aids to long-term
efficiency.

- But environmental considerations can affect Bank policy. The
Bank recently withdrew support for a cattle-ranching project in
the Vaupes region of Colombia after surveys showed the area's
forest soils to be nutrient-poor and incapable of supporting
large-scale development.

- Other aid and finance institutions have become increasingly
environment conscious. In some cases, such as an irrigation
project in Panama recently funded for $10 million by USAID, loans
are made contingent on the introduction of watershed management
or some other environmentally sound practice.

- The Inter-American Development Bank insisted on and funded the
establishment of a national forest service before approving a
tree plantation loan in Uruguay.

- In February 1980 the World Bank and the six major regional
development banks (Arab, Asian, European, Inter-American, African
and Caribbean), along with the Organization of American States
and the UN Development Programme, signed a Declaration on the
Environment.

- The document has no binding power, but the signatories agreed
to institute procedures for systematic evaluation of the
environmental consequences of their development activities and to
give technical and financial support to projects designed to
safeguard or improve the environment.

- There has been a major controversy over the effect of the World
Bank's policies on tribal peoples.

- Between 1969 and 1978, according to an internal Bank document,
over 170,000 members of tribal groups were forced to move so that
their lands could be flooded by Bank-financed hydroelectric
schemes.

- Following severe criticisms, the World Bank report "Economic
Development and Tribal Peoples: Human Ecologic Considerations"
was published in July 1981 by the organisation's Office of
Environmental Affairs. It recognised that "Bank-financed
development projects...will unavoidably impact (on the world's
200 million tribal peoples) often in undesirable ways". It also
admitted that the problems arising from such developments had
previously been played down.

- The report gives examples of projects seriously affected by
failure to take into account the needs, or occasionally the
existence, of tribal peoples, and points out that such peoples
"understand ecological inter-relations of the various components
of their resource base better than do most modern foresters,
biologists, agronomists and ecologists".

- The document has been criticised as ambiguous, and as unlikely
to be implemented. Nevertheless, it represents the most open and

authoritative admission of the problems yet made by any
development institution. It may help change attitudes towards
tribal groups, and towards the advisability of taking their needs
into account when formulating development schemes.

The plan that failed

In 1980 the UNEP Governing Council decided to back, along with
FAO and UNESCO, the preparation of a global "plan of action for
the wise management of tropical forests" which UNEP hoped to
publish in mid-1982 to coincide with the 10th anniversary of the
UN Conference on the Human Environment in Stockholm.

- In working papers the agencies repeatedly emphasised that the
plan was not "intended to infringe on the sovereign right of
states to decide on their own responsibilities in relation to
tropical forests. Its aim was, rather, to recommend an
appropriate division of responsibilities between governments and
international organisations."

- The first expert committee meeting on the plan (Nairobi, 1980)
also recognised "that the primary responsibility for use of
tropical forest resources to ensure the optimum flow of goods and
services now and in future rests with governments, assisted where
appropriate by international and non-governmental organisations,
and technical and scientific, and economic and cooperation
agencies".

- The second experts meeting in early 1982 in Rome was attended
by 34 experts from 21 nations. Such key tropical forest nations
as Brazil, Zaire, Colombia, Venezuela and Burma did not send
representatives. However, such "Northern" countries as Denmark,
France, West Germany, Japan, Netherlands, Norway, Britain and the
United States were represented.

- Despite the emphasis on national sovereignty and the realising
of the economic potential of forests, the meeting floundered over
the issue of sovereignty. All hope of a global plan was abandoned
and instead the experts identified 30 "elements" of action -
calls for more education, awareness raising, institution
building, resource management, etc - and left it to the
individual nations to assign priorities to the various elements.

- Tom Stoel, of the Natural Resources Defense Council in the
United States, said the Rome meeting "wound up reiterating the
conclusions of the first 1980 Nairobi meeting in a less coherent
way". It gave vague responsibilities for international tropical
forest work to the existing FAO Committee on Forestry Development
in the Tropics, provided that the committee's charter was changed
to give UNEP and UNESCO equal standing with FAO.

- "The tropical forest nations seem to want things to continue as
they are. They are not ready for change. The meeting was a great
disappointment", said an expert who attended.

- The discussion document for that meeting had said that the immediate cause of the "serious situation of the world's tropical forests" was "the poverty of the neighbouring populations. Tropical forest resource management must, therefore, include among its prime objectives that of alleviating rural poverty."

- However, as regards financing, the discussion paper could only offer the hope that money would come from "international banks such as the World Bank, regional banks and bilateral sources".

- It concluded with the hint that the nations with tropical forests must begin action themselves before they can expect much help: "finance may become more readily available if in developing countries in the tropics political decisions to promote the management of tropical forest resources are adopted and converted into practical action. This would stimulate the international community to provide much greater financial assistance than it has in the past".

A tropical hardwood log leaves western Brazil. As Asian forests are logged out, timber companies are beginning to turn to the TMF forests of Latin America. (see page 31)

Photo: *Marcos Santilli/Earthscan*

A vision of the future? A Nambiquara Indian of western Brazil standing amid the wasteland left by a highway which runs through several of her tribe's villages.   *(see page 23)*

Photo: Marcos Santilli/Earthscan

--oooOooo--

EARTHSCAN PAPERBACKS

<u>Drugs and the Third World</u> by Anil Agarwal (Out of print, but available in A4 lithoprinted format)
70 pages   £2.50/$6.25

<u>A Village in a Million</u> by Sumi Chauhan
ISBN No 0-905347-08-0
22 pages   £2.00/$5.00

<u>Climate and Mankind</u> by John Gribbin
ISBN No 0-905347-12-9
56 pages   £2.00/$5.00

<u>International Trade in Wildlife</u>
by Tim Inskipp and Sue Wells
ISBN No 0-905347-11-0
104 pages   £2.00/$5.00

<u>Antarctica and its Resources</u>
by Barbara Mitchell and Jon Tinker
ISBN No 0-905347-13-7
98 pages   £2.50/$6.25

<u>Mud, mud - The potential of earth-based materials for Third World housing</u> by Anil Agarwal
ISBN No 0-905347-18-8
100 pages   £2.50/$6.25

<u>New and Renewable Energies 1</u> (solar, biomass)
edited by Jon Tinker
ISBN No 0-905347-21-8
44 pages   £2.50/$6.25

<u>New and Renewable Energies 2</u> (others)
edited by Jon Tinker
ISBN No 0-905347-22-6
52 pages   £2.50/$6.25

<u>Water, Sanitation, Health - for All?</u> by Anil Agarwal,
James Kimondo, Gloria Moreno and Jon Tinker
ISBN No 0-905347-27-7
146 pages   £3.00/$7.00

<u>Carbon Dioxide, Climate and Man</u> by John Gribbin
ISBN No 0-905347-28-5
64 pages   £2.50/$6.25

<u>Fuel Alcohol - Energy and Environment in a Hungry World</u> by Bill Kovarik
ISBN No 0-905347-29-3
75 pages   £3.00/$7.00

<u>Stockholm Plus Ten - Promises, Promises?</u>
<u>The dedade since the 1972 UN Environment Conference</u>
by Robin Clarke and Lloyd Timberlake
ISBN No 0-905347-30-7
76 pages   £3.00/$7.00

All Earthscan paperbacks can be obtained from:
Earthscan, 10 Percy Street, London W1P 0DR, UK.